TEN BILLION TOMORROWS

ALSO BY BRIAN CLEGG

Final Frontier

Extra Sensory

Gravity

How to Build a Time Machine

Armageddon Science

Before the Big Bang

Upgrade Me

Light Years

The God Effect

TEN BILLION TOMORROWS

HOW SCIENCE FICTION TECHNOLOGY BECAME REALITY AND SHAPES THE FUTURE

BRIAN CLEGG

St. Martin's Press New York

 Printed in the United States of America. For information, address St. Martin's Press, 175 Fifth Avenue, New York, N.Y. 10010.

www.stmartins.com

Library of Congress Cataloging-in-Publication Data

Clegg, Brian.
Ten billion tomorrows : how science fiction technology became reality and shapes the future / Brian Clegg.—First edition.
p. cm.
Included bibliographical references and index.
ISBN 978-1-250-05785-3 (hardcover)
ISBN 978-1-4668-6192-3 (e-book)
1. Science fiction—History and criticism. 2. Science fiction films—History and criticism. 3. Science fiction—Influence. 4. Literature and technology. 5. Science and civilization. 6. Science—Popular works. I. Title.
PN3433.6.C536 2015
809.3'8762—dc23

2015028643

Our books may be purchased in bulk for promotional, educational, or business use. Please contact your local bookseller or the Macmillan Corporate and Premium Sales Department at (800) 221-7945, extension 5442, or by e-mail at MacmillanSpecialMarkets@macmillan.com.

First Edition: December 2015

10 9 8 7 6 5 4 3 2 1

FOR GILLIAN, CHELSEA, AND REBECCA

CONTENTS

ACKNOWLEDGMENTS

My thanks to my editors, Michael Homler and Lauren Jablonski, and everyone at St. Martin's Press who have made the book possible.

Although I must thank the academic writers, like philosophy professor Nick Bostrom, who have thought long and hard about the implications that changes in technology could have on society, I particularly have to think of the giants of science fiction who inspired me to get into science in the first place and who provided the inspiration for most of these chapters.

This list does include modern writers like Adam Roberts and the late, much lamented, Iain M. Banks, but primarily the thanks have to go to the classic writers who brought the field from pulp nonsense to the inspiring fiction of thought and speculation. So the ones I really need to mention are the likes of Isaac Asimov, Robert Heinlein, Frederik Pohl, Cyril Kornbluth, Alfred Bester, John Wyndham, Ray Bradbury—and many more, who didn't just entertain but strived to make their readers think. And though TV and film science fiction is generally considered second rate by genre experts, it has had a huge cultural impact—and so I have to also include the makers of the likes of *Doctor Who, Star Trek, Star Wars,* and *The Matrix,* among others.

Many scientists owe their careers to the inspiration of science fiction—and the same goes for science writers. I would also like to particularly thank Dr. Peet Morris for the times that he has said, "Have you thought of?" and given me a whole new chapter to consider.

TEN BILLION TOMORROWS

1.

THROUGH A GLASS, DARKLY

It was November 1963 and the future was scary and wonderful in equal measures. Science fiction told me so.

The whole world, it seemed, was reeling after the events in Dallas on November 22, 1963. But for me, as a young child, it was the following day that was far more significant. Because that's when the British TV show *Doctor Who* began. Later, the stories would be bolstered by real science that was amazing enough to be fiction, the kind of wondrous information that filled Carl Sagan's *Cosmos* and astronomer Patrick Moore's show *The Sky at Night,* then taken to a whole new fictional level by *Star Trek.* Thanks to science fiction, I knew what the future would be like. Like Howard Carter peering into Tutankhamun's tomb in 1922, I could see "Wonderful things." And I devoured them wholesale.

Movies started to have an influence too, starting with old classics like *Forbidden Planet* and building to the awesome experience of first seeing *2001: A Space Odyssey.* Then the books took hold. The first book I bought with my own money was H. G. Wells's *The Time Machine.* I still have that battered paperback. As it happened, I was disappointed with this particular tale. I didn't like the book at age eleven—and I still find it tedious—but luckily, guided by my father, a lifelong SF enthusiast, visits to the library had by then already immersed me in science fiction adventures from the golden age. Asimov, Kornbluth, Pohl, Wyndham, Bester, Heinlein, Bradbury, and their ilk were my guides to a riotous, speculative future.

Before I go any further, in case you are science fiction fan, I ought to explain why I am going to occasionally use the term "sci-fi."

Traditionally, those who read and enjoy science fiction have only accepted the abbreviation SF. "Sci-fi," whether you pronounce it "sky fi" or "sy fi" has been frowned upon. (There was an attempt to take the term into the fold in an ironic way by pronouncing it "skiffy" and applying it just to TV and movie space opera.) "Sci-fi" was originally coined affectionately, probably by superfan Forrest J. Ackerman, to draw a parallel with "hi fi"—but in practice it has mostly been used by the media, and particularly by those who don't understand the genre. However, I think it fits well with some of the subjects of this book, capturing as it does a slightly dated, but excitable, view of the topic, seen not so much from the viewpoint of fandom as in the eyes of the normal world, a world that was increasingly influenced by science fiction's imaginings.

Back when I really had no idea what to call the genre, it was impossible not to get excited (and frightened) by the extremes of technology that we see in science fiction. Looking with mature eyes at the real future that came to pass, there's a tendency to be disappointed. We are well past 2001 now and we don't have regular flights to a Moon base. For that matter, we don't have flying cars and battles aren't fought with ray guns or light sabers. But that misses the point of science fiction and its relationship with the world.

It is true that, very occasionally, science fiction has predicted something that then really happened. We often see references to Arthur C. Clarke's very accurate 1945 prediction of the geostationary communication satellite. Unfortunately, though Clarke was an accomplished science fiction author, this was a nonfiction article for the magazine *Wireless World*. There were also "ion drives" that propelled spaceships by repelling electrically charged particles in science fiction long before they became relatively commonplace as the thrusters used on real space vessels. And then there was the one-man prediction factory that was Herbert George Wells.

Impressively dramatic, and militaristic, predictions in science fiction came from that great pioneer of the field. Wells described the use of tanks in battles in "The Land Ironclads," a story from 1903, thirteen years before the real tanks were first built. He denied that he had any great inventiveness in this, claiming that he merely adapted existing ideas like Leonardo da Vinci's conical human-powered wheeled-and-armed vehicle. Then Wells came up with the idea of dropping bombs from airplanes, with his appropriately titled *The War in the Air*, published in 1908. It's true that there had already been bombs dropped from balloons, but here, just five years after the Wright brothers' first powered flight, was some impressive insight into how this new technology could transform warfare.

However, it was with a little-known (and practically unreadable) 1914 book called *The World Set Free*, that Wells gave his most notable assessment of the future. Here, the remarkable imagination of this uninspiring-looking Englishman brings us a story of a future where artificial radioactivity is harnessed to produce electricity and eventually, in a war in 1956 between America, England, and France on one side and Germany and Austria on the other, the world comes to the use of terrible weapons based on nuclear energy, for which Wells invented the name "atomic bombs." That's enough to get the hairs on the back of your neck rising, especially bearing in mind that this premonition came from the man who wrote *The Time Machine,* which is written in the first person by the time traveler.

Just how remarkable *The World Set Free* is can be seen by putting the picture of the future it portrayed alongside a few realities from history. The concept of radioactivity only dated back to the turn of the century, and as late as 1933 Ernest Rutherford, a massive contributor to atomic theory, was quoted as saying, "The energy produced by the breaking down of the atom is a very poor kind of thing. Anyone who expects a source of power from the transformation of these atoms is talking moonshine."

In the next year Leo Szilard came up with the concept of a nuclear chain reaction that would make harnessing nuclear power possible; in 1942 Enrico Fermi produced the first working reactor under the bleachers of a disused football stadium in Chicago. Just three years later, on July 16, 1945, the first atomic bomb was exploded in the Trinity test. Realistically, the remarkable success Wells had does not require a time machine or clairvoyance. He combined some powerful insights with just getting lucky, as occasionally people must. When we look across the broad sweep of science fiction, far more of what is portrayed, if considered a prediction, came to nothing.

It surely is hard to imagine a more tangible hit for science fiction when looking into the future than *The World Set Free*—and Wells should certainly be recognized for his achievements. You may wonder why you may not have heard of this book, or at the very least why it isn't as well known as *The Time Machine* or *The War of the Worlds*? The fact is that, by comparison with Wells's literary masterpieces, *The World Set Free* is a very dull read. And this reflects the most important thing to realize about science fiction. It isn't futurology.

Science fiction does not set out to predict the future—instead it's about asking, "What if?" for all kinds of scenarios. It doesn't matter if those possible futures are likely to happen or not, as long as they are interesting. The aim is to portray the human reaction to new and interesting circumstances. If the writer happens to be lucky enough to hit on a match with what really takes place in the future, that's great—but it certainly isn't the point of the stories. In the two words "science fiction," the "fiction" part has to dominate, because unless the book is a good tale, it doesn't matter how interesting or surprising the science it contains is.

The two great forefathers of science fiction, Jules Verne and H. G. Wells, took very different approaches that illustrate the delicate balance between the realities of science and the dramatic requirements of fiction to perfection. Verne was, frankly, stuffily dismissive of his far younger British challenger, commenting:

> I do not see the possibility of comparison between his work and mine. We do not proceed in the same manner. It occurs to me that his stories do not repose on a very scientific basis. . . . I make use of physics. He invents. I go to the moon in a cannon-ball, discharged from a cannon. Here there is no invention. He goes to Mars in an airship, which he constructs of a metal which does not obey the law of gravitation. *Ça c'est très joli*—but show me this metal. Let him produce it.

As twenty-first century science fiction writer Adam Roberts has pointed out, Verne was not doing a good job of defending himself with these words. While it is true that Wells did make up the highly unlikely "cavorite," a material that was opaque to gravity, once he had that invention he then employed it logically and consistently. His story required a novelty outside of known science, but then pretty much stuck to what physics would predict for its use. Verne, by contrast, despite using technology that really did exist in a simpler form—a cannon—totally ignored physics in the way that he made use of that technology. It was perfectly well understood in his day that human beings couldn't possibly stand the g-force that they would experience in a projectile accelerated fast enough inside a gun barrel to escape Earth's gravity, however well cushioned they might be, and that they would end up squashed to a pulp.

A mirror effect to science fiction's ability to make predictions is when real life copies—or at least is inspired by—the fiction writer's art. This is, if anything, science fiction's real claim to fame and influence in the world. It is not that SF managed to predict the future, but that it was an inspiration to those who have made the future happen, both in terms of encouraging positive discoveries and warning about potential disasters. The literati may find it distasteful, but science fiction has had an influence on popular culture for as long as it has existed and has had far more impact on everyday life than literary fiction. The disdain with which literary novelists have typically treated science fiction may well emerge from jealously, because their beautifully crafted works often have a far smaller audience. Many scientists and engineers admit that they were fans of science fiction in their teens and it was this, in part, that inspired them to get involved in their profession, driven by the sense of wonder that comes from good SF.

The development of space travel was tied strongly into an archetypal fictional theme, pushed forward by science fiction dreams. When Wernher von Braun was developing the V-2 rockets that were used to attack England and Belgium during the Second World War, the military use of the technology was, to him, a mere distraction. In his mind, his efforts were always directed toward getting men into space. (This is not to belittle in any way the death and destruction caused by the V-2 program, but space travel was the reality of von Braun's inspiration.) Konstantin Tsiolkovsky, the Russian theorist who was arguably the father of practical rocketry, put as much effort into his science fiction writing as he did engineering.

The negative side of this inspirational connection running from fiction into the world has been the production of whole swathes of pseudoscience, driven by the imaginings of writers. Much of the flying saucer craze and the descriptions of alien visitors seem to have been inspired by their fictional equivalents. When science fiction writers described aliens as little green men, that's what people saw. When *The X-Files* and other shows and movies came up with big-eyed "gray" aliens that was what followed in alien sightings and abduction stories.

Arguably too, it was fiction like Robert Louis Stevenson's *The Strange Case of Dr. Jekyll and Mr. Hyde* that lay behind the work of Sigmund Freud (who came up with a similar concept of the human mind having a primitive, partially controlled part, the id, that corresponds closely to Stevenson's Hyde). Science now largely considers Freud's work as fiction itself, even though it is still often used as the justification for the methodology used in analysis and counseling. And, of course, the whole Scientology movement has its roots in the work of science fiction writer L. Ron Hubbard.

Even though science fiction isn't a mental time machine that can give us a peek into what is going to happen to us in the future, it inevitably places its characters in a landscape of strange science and innovative technology. This book celebrates the wonderful imagination of science fiction writers and the dramatic impact of real-world science and technology in those same areas, which can be every bit as remarkable as fiction expected it to be—and was sometimes inspired by that fiction—but often turns out to work and to be used in very different ways.

There is no other type of writing that can take the basic driver of all fiction—the way that the characters react to challenges and changes in the world and themselves—and give it license to cover so many different areas as does science fiction. With the imagination of SF writers at work, the landscape, the technology that surrounds us, even the very nature of human beings can be modified to try out and play "What if?" in astounding new possibilities.

Of course there has been plenty of trash written in the history of science fiction, and plenty that was just a reworking of an existing story in new, glossy surroundings. This is an approach that doesn't even have to produce bad science fiction—the classic SF movie *Forbidden Planet* is a great example, drawing heavily on Shakespeare's play *The Tempest* as source. But science fiction at its best has also given us the chance to think about and enjoy totally new challenges for human (and nonhuman) life to face.

Despite the caricatures and much of the sci-fi output of Hollywood, science fiction is not all about spaceships and ray guns. But there is no denying that part of its appeal (especially, perhaps, for the younger reader) is the wow factor of amazing imagined technology. It is fascinating to put the science fiction imaginings alongside what the real world has delivered, from artificial intelligence to the Matrix, both considering what has happened so far and looking forward to what today's science will make possible in the future. The results can be bizarre and fascinating—we might not have *Star Trek*'s terrifying Borg, but we do have remote-controlled beetles and home-kit cyborg cockroaches. The real thing often manages simultaneously to fall behind and to leap ahead of the science fiction equivalent.

Apple's iPhone voice interaction system Siri may be no intellectual match for the talking computer Hal in *2001: A Space Odyssey,* but Siri runs on a tiny phone, where Hal required a mainframe the size of a house.

I suppose, since I am using science fiction as a source and an inspiration, I ought to try to define what science fiction *is*. Fantasies about journeys into space or to strange lands (as featured in Jonathan Swift's 1726 book *Gulliver's Travels*) have been in existence for hundreds of years—in fact arguably Homer's *Odyssey,* the second oldest surviving piece of literature in the Western tradition is just such a fantasy. This is a parallel but separate strand from science fiction. While both involve a story based on "What if?," science fiction requires at least a hat tip toward what is physically possible, even if there are many bits of imaginary science like faster than light travel that are regularly invoked.

There was a strong science fiction element in gothic works like Mary Godwin's *Frankenstein* (she was yet to marry the romantic poet Shelley when she sketched out the story), but it was with the likes of H. G. Wells and Jules Verne that true science fiction came into being. Their stories were known as "scientific romances"—it was not until the 1920s and 1930s that the term "science fiction" itself was first used, whether in its present form or as "scientifiction," the uncomfortable mash-up devised by the SF pulp magazine pioneer Hugo Gernsback. He would define scientifiction as "a charming romance intermingled with scientific fact and prophetic vision . . . Not only do these amazing tales make tremendously interesting reading—they are also instructive. They supply knowledge . . ."

Gernsback had a stiff, old-fashioned educational idea of science fiction's role, but the pulp magazines soon transformed his worthy intentions into a wild flurry of entertainment, wonder, and horror—despite always keeping that touchstone of science.

A modern definition of SF might be something like "fiction in which science and technology is used as a setting in which to explore human (or nonhuman) behavior." The science is usually not an end in itself, though SF can certainly glory in the remarkable ideas it can present, and the technology that such original thinking can produce. At its heart, good science fiction is almost always about people.

Take *The Forever War,* Joe Haldeman's now largely forgotten counter to Robert Heinlein's gung-ho "grunts in space" novel, *Starship Troopers.* Haldeman's book, certainly featured some interesting science, exploring the way that time slows down when spaceships get near to light speed. But the heart of the story was the psychological impact on humans of going on missions deep into the galaxy where they would return to find everyone they knew and loved long dead. Even as hackneyed a space opera as *Battlestar Galactica,* the 1980s TV attempt to cash in on *Star Wars,* become a much more interesting story (arguably more interesting than *Star Wars* itself) in its twenty-first century reboot. In the new version, the strong backbone of the story line was not the spaceships and dogfights, even though they certainly featured, but the mental state of the key characters.

A cynic might say that the best definition of science fiction is: "What science fiction authors write." It is certainly true that when a literary fiction author ventures into SF, the outcome is often not called science fiction, because the genre has always been considered second class by those who move in literary circles. The term "speculative fiction" has sometimes been used instead, to emphasize the importance to SF of asking "What if?" and reflecting the way that it portrays the outcome of that exploration. Pure speculative fiction is probably a wider category though, as arguably it also takes in fantasy.

Look in the SF section in a general bookstore and you will

find that there are significantly more "swords and sorcery" books (think of the hugely successful *Game of Thrones,* or the great original *The Lord of the Rings*) than there are science fiction books. As we have seen, fantasy has always been lumped together with science fiction, but the dividing line for this book is whether or not something is even conceivably possible according to science as we know it. This "conceivably possible" can be stretched a long way. For instance, Einstein tells us we can't travel faster than light—but there are plenty of ways to bend the rules by warping space and time. And even some of the apparently inviolable laws of physics could be different in an alternate universe.

If the fiction is set in our world it needs to keep within those bounds to be true science fiction. Magic is unlikely to appear because it always presents problems for conservation of mass and energy. It's fine for science that is far beyond the characters' imagining to be seen by *them* as magic. And there is nothing wrong with dreaming up variants of science that have yet to be discovered and could still fit with what we know. But it is important not to go too far out on a limb. So despite there being some remarkable science in this book, and technology that it is hard to imagine ever becoming real, don't expect to come across a flying dragon the size of a blue whale or spiders the size of elephants, because physics just won't allow it.

As you explore *Ten Billion Tomorrows,* the chances are that you will be tempted to point out a favorite piece of technology or a pet science fiction author that I've missed. I have made no attempt to be comprehensive. Rather than struggle to list every possible piece of science fiction technology and detail its real-world equivalent, ending up with an encyclopedia-like collection, each chapter is built around a specific example, but then takes the opportunity to bring in a range of other devices and technologies that link to the subject. While the chapter that features

Iron Man majors on the development of the rocket belt, it also brings in other kinds of personal flying transport and the development of exoskeletons. This does mean that there are some sad exceptions, but the line had to be drawn somewhere if this was not to become a list book.

It might feel as if the result of my efforts is a rosy, technophile's view of the world. We know that technology can bring negatives as well as positives, and sometimes this will emerge, as, for example, when we look at the difficulties of bringing a powerful artificial intelligence under control. Yet I make no apology for being broadly positive about the way science and technology has influenced our lives—because overall the influence of this most remarkable development of human endeavor has been startlingly good.

If you look back before the benefits of twentieth-century science and technology, for all but the privileged few, life was thoroughly miserable. I don't deny that there are still many millions today who suffer this kind of lifestyle, but it is only because of modern developments that many of us whose ancestors were servants or farm laborers, people who spent long days barely making enough to live, now enjoy a remarkably free lifestyle, able to cross the world in hours physically, or in seconds electronically. In just over a hundred years, medicine has changed from guesswork to science, extending lives and preventing many illnesses. We can communicate, work, be educated, and have fun in far many more ways than our ancestors—and unlike them, most of our children don't die before they reach adulthood.

It's easy to moan and be pessimistic about the doom that science and technology can bring us, but I think in being inspired by the science fiction antecedents we can take the liberty of enjoying the sense of wonder, the real youthful joy that the capabilities of the fictional inventions brought—and to allow ourselves the liberty to have a similar sense of wonder and thrill from the truly remarkable achievements of real-world scientists and engineers.

It's easy to forget how much has changed in existing lifetimes. Back in my childhood there were no cell phones, there were no color TVs, and in the UK we only had two channels to watch. We didn't get a fridge until I was ten or central heating until I was eleven. I had friends at primary school who had no real bathroom in their house, still using a tin bath in front of the fire. Science fiction's glittering future presented a different world, and though what we have now is an alternate reality to the visions of the science fiction writers, it still has much of wonder to offer.

Where to start? There are so many places. But to emphasize that all the best presentations of science fiction ideas didn't come in the golden age of written SF between the 1940s and 1960s, let's start with an iconic decision:

Blue pill or red pill? What's it to be?

2.

BLUE PILL OR RED PILL?

There are some movies that bring back the teen joy of discovering science fiction for the first time perfectly. They are the ones where you sit in the theater and think "Oh, wow!"

Such sci-fi gems don't have to be the best films ever made or even to have the most effective scripts—instead they work on a level that is both visceral and captivating. It's hard now to recall just how much this was the case with *The Matrix,* where that "blue pill or red pill" choice featured so memorably, back in 1999. I may have been an adult, but *The Matrix* appealed to the inner teenager that most of us keep bundled inside. Many viewers fell a little in love with this movie for all its flaws. As soon as Neo was rushing through the building, directed by the mysterious voice on the phone, and got that "No, the other left!" instruction, we were hooked. It gave that same out-of-control thrill that we experience when riding a roller coaster.

The year 1999 may seem a long way back. (Can't believe that *The Matrix* was so long ago? Take another look at the "hi-tech" cellphones that are central to the plot. They are Nokia flip phones with a tiny monochrome screen.) It was, after all, the last century—the previous millennium—but, strangely, this wasn't the first time the Matrix cropped up. I had come across the concept once before. There is no suggestion that the Wachowski siblings, who wrote and directed the movie, didn't come up with the idea entirely independently, but by a bizarre coincidence, the long-running science fiction TV show *Doctor Who* included a computer-based virtual reality universe called "the Matrix," in a 1976 episode. Featuring in "The Deadly Assassin," the Matrix was a computer system that provided a virtual world in which real people were immersed by wearing special helmets. In the electronic world, their realistic avatars could battle each other—and if they were killed there, the real person who was hooked in would die as well. Ring any bells?

Back in 1976 when the *Doctor Who* episode was shown, the term "cyberspace" had yet to be used—it would still be six years before William Gibson brought it to a wide public in a story written for *Omni* magazine. Though the ARPANET, the antiquated predecessor of the Internet, was already in place, in was at the time just a means to remotely log in to other computers in universities and military institutions. The true Internet was arguably not started until 1982 (alongside that *Omni* article, with reality running parallel to fiction) while Tim Berners-Lee did not start work on the World Wide Web at CERN until 1990.

By coincidence, 1976 was also the year when a true computer-based virtual world came to life. It was then that American computer engineer Will Crowther, who was working on ARPANET at the time, had an idea that would capture the hearts and minds of computer enthusiasts—me included. Crowther was a fan of the fantasy pen-and-paper role-play game *Dungeons & Dragons,* which came out in 1974. He was also having family issues at the time. A caver, he had mapped out a real cave system on an early computer, and used the idea (if not the actual map) as an inspiration when he wanted to have something to play with his daughters after his marriage breakup. So he put together a simple game that used the computer's ability to respond to a series of text commands to build a virtual world, which would later be improved on by graduate student Don Woods, who strengthened the fantasy elements in the game.

When playing, the early gamers would be told their position in a series of linked caves and they could ask to move in different directions. They might discover swords or treasure—or, for that matter, deadly monsters—all of which were summed up with a few, tightly conceived words. Any pictures were in the imagination of the player. Crowther called his game *Adventure,* set in Colossal Cave. This wasn't the first game to make use of computer power. The stand-alone tennis game *Pong,* running on TV sets with a simple computerized controller to produce the signal, came onto the market in 1972. But Crowther's *Adventure* was the first computerized adventure game (giving the name to the genre).

In 1976 I moved from the bustling world of Cambridge to the isolated campus of Lancaster University in the north of England to take my masters degree. Until then, my use of computers had been limited to running simple programs written in languages like Fortran on a stack of punched cards. But Lancaster had a secret computing weapon in George 3, a computer-operating system running on an already antiquated 1900 series mainframe from the now long-defunct ICL company, an operating system that transformed the way that the user interacted with the computer. Instead of punching a set of cards, feeding them through a reader, and waiting for the output to churn out on a line printer, users communicated with George 3 using teletypes, electronic typewriters that could both take input from a keyboard and respond by typing controlled directly from the computer. Suddenly it was possible to have a conversation with a computer in real time.

I made a small amount of use of George 3 for my coursework. But by far the most frequent command I would type after logging in to the system was ADVENT, the command to run the game *Adventure,* which had been ported from its original implementation on a DEC PDP-10. I would play long into the night under the stark fluorescent lights of the computer lab, immersed in a world that could only be accessed via the imagination and the clattering printhead of the teletype:

YOU ARE STANDING AT THE END OF A ROAD BEFORE A SMALL BRICK BUILDING. AROUND YOU IS A FOREST. A SMALL STREAM FLOWS OUT OF THE BUILDING AND DOWN A GULLEY.

When in response I typed GO IN, the system would respond:

YOU ARE INSIDE A BUILDING. A WELL HOUSE FOR A LARGE SPRING.
THERE ARE SOME KEYS ON THE GROUND HERE.
THERE IS A SHINY BRASS LAMP NEARBY.
THERE IS FOOD HERE.
THERE IS A BOTTLE OF WATER HERE.

The adventure had begun. I genuinely can still feel the hairs standing up on my arms when I read those nostalgic words again. There had never been anything quite like it. It was a whole world that you interacted with through text—these text-adventure games have become known as interactive fiction, and the name fits well. Perhaps most extraordinarily back then, it was a computer program that responded to ordinary words, rather than the terse instructions of a command language. The game would stay with me in 1977 when I moved to work at British Airways, where we had a PDP-10 and were able to play the original in all its glory. There too came another step toward the world of *The Matrix* in the ability to link people together as they played. *Adventure* was fun to play on your own, but the Star Trek game *DECWAR* was something entirely different.

This was a true multiplayer game. All the players were equipped with teletypes (or for a few lucky players a VDU, a display screen involving no paper and printhead), and could be in different rooms or different buildings. Each player joined the game as a member of the Federation or a Klingon. As your ship flew around, you could detect other ships, communicate with them, or attack them with phasers and photon torpedoes. Those of us who played regularly could annihilate a new player before they even worked out how to set a course. It was one more step to bringing a science fiction world into a state of virtual reality.

Text adventures and computer-based Star Trek battles were a very cerebral form of virtual reality. With time, the adventures gained static illustrations, painfully loaded from cassette tapes, and even some crude graphics that allowed for a changing view around the player, but in 1987, a young computer enthusiast got the jolt that would inspire virtual reality developments that went far beyond the lure of "You are standing at the end of a road before a small brick building." That computer enthusiast's name was John Carmack, and he would be responsible for the technology behind three computer games that literally rewrote the landscape of the gaming world. What inspired Carmack, was the holodeck on the shiny reboot of *Star Trek* called *Star Trek: The Next Generation.*

We will return to the holodeck in chapter 4, as the technology behind it comes from a whole different technological arena, but the concept stuck with Carmack. What the teenager wanted to do was create a world on the computer screen that would fully immerse the player, making them feel as if they were part of the action. At the time, computer graphics, particularly on the then relatively new IBM PC, were dire. Yet the first major game produced by id Software, the company formed by Carmack and game designer John Romero, *Wolfenstein 3D,* managed to do remarkable things on a PC.

Admittedly the game was visually blocky and limited in its use of color, but by making use of every software trick in the book, and many that weren't until he wrote them, Carmack managed to make the journey through the castle taken by the player, who saw a first-person view, smoother and more convincing than anything that had ever come before. Another company joined the race to raise the bar in 1993 with an immersive adventure game called *Myst,* initially written for the new color Mac computers. Underlying the game was a very similar kind of experience to that in *Adventure.* The player moved around from place to place searching for clues and objects. But here, for the first time, the visual experience was startlingly like an idealized form of reality. What was seen on the screen was not blocky graphics, but exquisitely detailed computer-generated images.

Adventure-game players were wowed by *Myst,* which was soon ported to the PC, a platform with increasingly effective graphics, where it became the bestselling game for a number of years. But John Carmack was not impressed by the rival product. *Myst*'s graphics were, indeed, glorious, but the game lacked the fundamental requirement needed to make it anything close to virtual reality. You couldn't move and look wherever you liked. *Myst* had a series of set locations that you could jump between. At each location you could look around you, and interact with objects and people, but any attempt to move elsewhere sent you to the next location, which could be some distance away. Like *Adventure,* it was a series of unconnected nodes in a network, not a space that it was possible to actively move through.

For Cormack, such freedom of virtual movement was the essential in taking a step toward the holodeck, and he would take the approach to a whole new level in the follow-up to *Wolfenstein,* called *Doom.* The visuals might not have been as lush and sophisticated as those in *Myst,* but this was a world that the player could move through, interacting with everything around him (or her) with no artificial jumps from location to location. This title, and its more sophisticated successor, *Quake,* set the stage for all the first-person games that have come since—games that have dominated the modern console market as well as the PC. To a player, the games might be all about fighting and shooting enemies, but the challenge to Carmack was to make the experience of moving around the virtual world as much as possible like reality—to re-create the visual perfection of the holodeck on a computer screen.

It is interesting that Carmack is, at the time of writing, the chief technology officer for Oculus VR, the leading developer of virtual-reality goggles. These are part of the extended virtual-reality world that attempts to go beyond what is possible when we simply look at a screen. It is perfectly possible to get immersed in a game (or a movie) and to forget at some level that you are not present—but physical enhancements like goggles, which give much more feeling of presence by changing the view as your head moves, or gloves and movement sensors that enable your character in the virtual world to move as you do, take that immersion far deeper.

Strap-on technology can go so far, but for the inhabitants of the Matrix—and many other science fiction visions of virtual reality, occupants of that 1980s William Gibson term "cyberspace"—

the virtual world and the real one become indistinguishable by bypassing the sensory organs and going straight for the brain. If there is a distinction, it tends to be that the virtual world is more colorful and engaging than reality. To make this possible there is only so much that can be done by strapping on equipment. It isn't the same as reality. But to the occupants of the Matrix, with the exception of the occasional glitch like a reoccurring cat, the experience was indistinguishable from the brain's interaction with a physical world. In the Matrix and its equivalents, instead of interpreting signals from the sensory organs (and these could be touch and movement as much as sight and sound), the information is pumped straight to the brain. The experience could then be as real as the technology could make it.

With today's technology we are some distance from what is possible in the Matrix in two senses. One break point is the interface and the other is the level of detail that can be created. As we will see in chapter 15 on cyborgs, there has already been work done sending signals into electrodes inserted into the brain to attempt to restore sight. But any operation on the brain is nontrivial, and unlikely to be undertaken just to provide the kind of interface depicted by the jack plugs used in *The Matrix*. There, the sockets have been inserted involuntarily by the machine race that has taken over the world. But if we decided to go down this route in reality, we would be making a conscious decision to undertake risky surgery for a form of entertainment—not something many would consent to, even to produce a near-perfect virtual world.

The ideal would be to move away from inserting electrodes into the brain—which, apart from the dangers of the operation itself, has proved to be a short-term solution as the electrodes are eventually rejected. Instead we need to find ways to be able to monitor and interact with the brain from an external device.

Even if the connection *had* to be made physically, in reality there would be no external socket as in *The Matrix* as the places where it is linked to the flesh provide too easy a source of infection—instead there would be some sort of wireless link to a subcutaneous device. Making a real link to the Matrix would be more likely to use brain Bluetooth than a jack.

The dangers of intrusive brain connections are so great that there almost has to be a form of indirect external connection, and crude versions do exist. Picking up information from the brain would probably use a more advanced version of an EEG (electroencephalograph), which uses a set of electrodes attached to the scalp (or just resting on the head in a cap) to detect signals from within the brain. Every time a neuron in the brain fires, tiny electrical charges are produced, creating electromagnetic waves that induce currents in the electrodes. It's like tuning in to a massive collection of miniature, short-range radio transmitters.

The trouble with this approach is just what you'd expect when attempting to distinguish the signal from billions of transmitters, all located within a few inches of each other. There are so many neurons in the brain that it is currently impractical for an EEG to detect anything other than the average output across many millions of different cells. The result, compared with the precision that direct electrodes can provide, is a blurry, limited readout that is easily corrupted by the output from other brain activity. Even so, such EEG helmets are beginning to improve in sophistication, and there is no doubt that with time, our scanning methods will get more sophisticated and be able to accept detailed commands from the brain.

Picking up instructions from the brain is one part of the fully immersive virtual-reality world. There is no point being in it if you can't control your virtual body. But the other requirement is to reverse the effect of the EEG helmet and pump information into the neurons, stimulating the same responses as would be generated by sensory inputs. To do this is further down the line than a high-resolution EEG. Some work has been done using transcranial magnetic stimulation (TMS), where electromagnetic induction is used to produce small currents in the brain as a result of fluctuating magnetic fields on the outside. This has a number of medical applications, but as yet is unable to generate a stimulus with anywhere near sufficient accuracy to produce a controlled sensory experience. As yet a workable version of this technology seems a long way off, though often breakthroughs in fields like this come when a whole new approach, as yet unenvisaged, is taken.

The challenges are big enough in terms of providing an effective interface to the brain, but the problems of recreating the Matrix don't stop there. Because it is one thing to be able to stimulate the visual cortex and another to be able to generate an image (and all the other sensory stimuli) at will. First we need to generate a view of the world that is of the resolution of the human eye and that changes almost immediately as we move around. This is the challenge John Carmack faced, magnified many times over. While it's true that modern movies contain CGI worlds that are pretty well indistinguishable from reality on the big screen, these images are not created in real time, but have to be laboriously built up, frame by frame. We've a way to go before a full resolution human's-eye view can be generated with the fluidity of a game (though this is only a matter of time and computing power).

Once that is achieved, we have to match those visuals (and other senses) against the required neural stimulation. In 1999 a team at Harvard led by Garrett Stanley took signals from a cat's retina and used a computer to convert these electrical impulses into a TV picture. They were able to present a crude representation of what the cat saw on the screen. This was a kind of reverse of what's needed—if it's possible to map the incoming signals to the right places, in theory that same image could be played on a cat's brain. But there is a long way to go indeed until we can experience a computer-generated world, like those in the movie *Avatar,* played out our brain's artificial screen that is usually populated by the output of our sensory organs.

In practise it *is* possible to have the level of believability that the Matrix provided in a virtual reality environment in the real world right now, and has been for many thousands of years. There's a natural virtual reality that most of us experience every night that we call dreaming. Unlike the Matrix, a dream doesn't have to cover up glitches particularly well, because in the dream state the mental checks and balances that allow us to spot that there is something wrong with our environment are disengaged. In a dream we can transition straight from one place to another, without feeling that anything odd has happened. In a sense, dreams are more like *Myst* than *Doom* (which is probably just as well).

Because dreams already exist, if largely in an uncontrolled fashion, a more realistic portrayal of virtual reality in science fiction may be the one that comes in the 2010 movie *Inception.* Here, the central characters enter another person's dreams in order to garner information or even (the "inception" of the title) in an attempt to influence waking behavior. While a lot of the truly impressive aspects of *Inception* are down to either the spectacular graphics, as when the streets of Paris fold in on themselves, and the mind-twisting possibilities of dealing with a dream within a dream within a dream within a dream, the idea that a virtual-reality world could take place within another person's mind has distinct possibilities. It is easier to imagine that complexity of setting and population taking part in a brain than any current computer.

Although much of the movie *The Matrix* takes place inside the artificial virtual-reality world, there are also scenes in the "real" world, a place that has been ravaged by the machines and where humans hide in hovercraft and subterranean vaults. On these ships, before entering the Matrix, it's not uncommon for participants to take on a new expertise downloaded from a computer disk. They may need to know how handle particular weapons, or to be able to fly a helicopter. No problem—slot in the disc, download the expertise, and within minutes they are an expert.

The whole, tedious business of learning has seemed for many years to be far too much effort and science fiction has regularly tried to come up with ways to get information into our memories without all the hard work and (let's be honest) the boredom we often associate with schooling. The method in *The Matrix* seems to assume that if we can access the brain in sufficient detail to set up the experience of being in a world, we can also transfer memories of how to undertake a task to the appropriate parts of the brain, but as we will see, this is probably a significantly more difficult task than simply projecting a virtual reality onto the parts of the brain handling the senses.

Elsewhere in science fiction, there are two broad approaches to the learning processes of the future, accelerated learning or reprogramming the brain, as one would transfer a backup of a hard disk from one computer to another, after which the second computer "remembers" how to do everything the first one did. A good example of the first of these methods came in the Cities in Flight series, begun in the 1950s by James Blish. In *A Life for the Stars,* the young protagonist Chris deFord finds himself accidentally taken off the Earth on the city of Scranton, one of the later cities to leave the wrecked planet using the new spindizzy drives that allows a whole city to act as a vast spaceship. DeFord ends up transferring to New York, where he is given the option of taking on the required education to become a citizen.

This education process is described as "accelerated schooling." The students lie on couches in a gray room with helmets containing thousands of electrical contacts on their heads, while a gray gas fills the room. In a sleep-like state a "torrent of facts" is poured into the participants' brains. This is not the trivial "slot in a memory card and absorb the program" of *The Matrix*—it is long, hard, draining work—some of the students have to give up after suffering fits. But the result is an education that can be crammed into months rather than years.

The second approach, a "download a brain" method is taken in the short-lived series *Dollhouse* by *Buffy the Vampire Slayer* creator Joss Whedon. The "dolls" in the title are individuals who have contracted to give up five years of their life in return for riches in a new future. During that period, they are wiped of their own memories and sent on missions where they are programmed by giving them the memories and personalities of one or more individuals, giving them expertise that could range from the skills of a courtesan to a hyper-efficient assassin. These memories are inserted into the doll's brain using a chair with some kind of remote electromagnetic stimulus, transferring information stored on what appear to be computer hard drives (referred to in the show as "wedges").

The *Dollhouse* approach, which bears a resemblance to a whole-brain version of the learning process in *The Matrix*, seems to underestimate the complexity of what's going on inside a human skull. The number of potential connections of all the neurons in the brain provides a combinatorial explosion that would require every atom in the universe if we were to try to map out every possible combination. Of course, if the brain can store the data, so can an electronic device, but even in the actual connections in any particular brain, we are talking far more storage than is feasible in a compact device at the moment.

In a sense, the *Dollhouse* approach is more sensible than that in *The Matrix*, as it doesn't require the programmer to pinpoint just where the expertise is recorded in order to be able to reproduce it. The brain is not like a computer disk. It does not store memories in one place, but spreads them throughout the brain, using a series of links and branches. What's more, there is more than one kind of memory. Facts seem to be processed by the hippocampus and spread through the cortex. But these so-called "declarative," high-level memories are totally different from the memory of how to do something, the low-level, procedural memories, which reside in the "old" part of the brain, the cerebellum, located near the spinal cord.

When we learn how to perform a task we start off using the factual part of memory—we have to think through what we are doing step-by-step, and this takes time. This is why we are so bad at driving or playing the piano when we first start. We have to consciously make the decisions of what actions to take in what order. But over time, the abilities become stored in the cerebellum. This procedural memory does not require conscious control, but connects quickly to the nervous system to initiate actions without thinking things through.

If, like me, you can touch-type, it is easy to see this distinction in action. Let's say I wanted to type a letter C. I just did it when typing that sentence without looking at the keyboard. All I needed to know was that I wanted a C and I hit the correct key. But ask me where the C is on the keyboard, and I can't tell you. I never had training to touch-type, I just picked it up, and I don't have any mental picture stored of the keyboard layout. (I know the top row is QWERTYUIOP, but that's all.) I just think C and it happens. The procedural memory does the job without bringing conscious declarative information into play.

This is great for actually getting things done, but it means that the idea of somehow being able to extract the expertise of, say, how to fly a particular helicopter, and inject it into someone's brain as happens in *The Matrix* is highly unlikely. Not only would the software developer have to locate all the different places in the brain where "how to fly a helicopter" is located, the program would somehow have to pull together the factual information on what the different displays means and how to use them, and the procedural data that allows a pilot to fly without thinking about every action. It's a very big ask.

Of course, as yet we haven't even come close. The technology to achieve what we see in *The Matrix,* or in a *Dollhouse*-style brain upload is not yet imaginable. There have, though, been many attempts over the years to achieve something like the accelerated schooling portrayed in *A Life for the Stars,* and the chances are that Blish was basing his imaginary technique on what was thought at the time he wrote the book to be an effective technique of unconscious teaching called hypnopedia or sleep learning. Unfortunately, in reality, the hype exceeded the capability.

The idea seems to have been that sleep was related to being in a hypnotic state, where the mind was open to suggestion (overlooking the fact that, even if this vast oversimplification was true, there was no evidence that being hypnotized helped people to retain information for very long). So, it was thought, all that was necessary to push the facts into long-term memory was to talk to someone while they were asleep or, more practically, to place a loudspeaker under a sleeper's pillow and play back information during the night as they slept.

This idea (and a range of products that have been sold over the years based on it, including smartphone apps available today) seems to have been based on pure theorizing without any practical experimental verification, and when experiments were undertaken as early as 1956, the researchers discovered that the only retention occurred when the sounds were played during the period when an EEG machine recorded alpha wave activity, which showed that the subjects were waking—in effect they were simply remembering something they had heard when they were awake—and this gave no benefit over the limited learning ability provided by just hearing something once.

Just because the sleep-learning machines of science fiction don't translate into reality doesn't mean that it isn't possible to make use of sleep to enhance retention of memories. You can't actually learn while you sleep, but it does seem possible to enhance sleep's role in arranging and consolidating memories to make it easier to fix information learned earlier in the previous day in long-term memory. It was already known that the sense of smell is the easiest to stimulate while sleeping, and experiments have shown that if someone experiences a particular, distinct smell while learning something, and they are then exposed to that same smell while sleeping that night, their recall of the information is better in the morning. Brain scans show a better connection between the hippocampus, which directs memory storage and other parts of the brain—the smell seems to have helped lock the memory into place.

Similar effects have been generated by playing music during learning and repeating the same music during sleep that night. Something perhaps closer still to the science fiction approach was an experiment that used external electrical stimulation of the brain to amplify the waves of electrical activity within the brain that accompany memory consolidation. Again there was some evidence of improvement in memories being fixed for later recall. It's early days, and the approach is very different from the fictional version, but it looks like we can enhance sleep's role in storing memories away.

Another feature of the "real" world of *The Matrix* movie that is more technically simple than memory manipulation is the EMP or electromagnetic pulse. When the menacing, jellyfish-like flying robots are advancing on one of the human-controlled hovercraft, an EMP is used to destroy all the nearby robots, frying their electronic circuitry with a huge spike of electromagnetism. (This makes for a good plot device, as the same pulse would also fry the hovercraft's circuits, so it has to be shut down before using the EMP. In reality, the circuits would probably be fried whether or not they were shut down, but this is a story.) Making use of an EMP this way is, in principle, a very simple technology and it is something that happens regularly as a result of natural electromagnetic pulses, which are usually caused by lightning.

A burst of lightning is an electrical discharge on a truly massive scale. The power in a lightning bolt is phenomenal. To run a 100-watt bulb for a second takes 100 joules of energy. A typical lightning flash carries over half a billion joules—that's the output of a medium-sized power station run for a second, all released in a tiny amount of time. The energy in the discharge blasts air molecules into frantic activity. The temperature in the vicinity of a lightning bolt can reach between 20,000 and 30,000 kelvins (53,540 degrees Farenheit) far hotter than the surface of the Sun. As the air expands violently away as a result of this sudden increase in temperature, it produces a shock wave that we hear as thunder. Thunder is, in effect, air being catapulted away from a lightning flash.

The visible flare of light that typifies lightning and seems to split the sky is not the electrical current. The reason a heated material glows is that electrons in its atoms have been given extra energy, then dropped back to a lower level, releasing the extra energy as a photon of light. It is the energized atoms in the air that provide the distinctive glow of a lightning bolt. And the light we see is just a tiny fraction of the entire electromagnetic spectrum, which stretches all the way from low frequency radio waves through microwaves, infrared, visible, ultraviolet, X-rays, and gamma rays. Many lightning bolts emit electromagnetic radiation all the way up to X-rays.

When an electrical current flows through a conductor it will induce a current in any nearby conductor. That's how a transformer works—there is no direct connection between the two circuits inside it, but one winding induces a current in the other. As the strength of current changes (a transformer uses AC) this produces a varying magnetic field. And a varying magnetic field produces an electrical current in the second winding. When a lightning bolt occurs the current surges from nothing to as much as 30,000 amps. This produces a huge magnetic pulse, which will induce a strong, if transient, electrical current in any nearby piece of metal, and this means that sensitive electronic equipment like computers and Internet routers can be knocked out permanently by the electrical pulse that flows through their cables and internal circuitry. It's a good idea to link equipment like this to the mains through "surge protectors," which iron out pulses and reduce the likelihood of damage.

On a far larger scale, the Sun is capable of generating a tremendous EMP that could knock out virtually all of the Earth's electronics, and even damage less-sophisticated electrical devices like electric motors. On a regular, though thankfully not too frequent basis, the Sun undergoes solar flares or coronal mass ejections, where a huge storm on the surface of our neighborhood star propels large amounts of charged material out into space, with the potential of producing a huge EMP on the Earth.

Back in 2012, on July 23, there was concern among those in the know about the impact of a huge solar storm. As it happens, the ejection occurred in a direction away from the Earth, but had it come just a few days earlier, we would have felt its full impact. If the flare's output had bathed the Earth, we would probably have lost *all* our satellites for good. Not just telescopes, like the Hubble Space observatory, but GPS satellites, weather satellites, and communication satellites would all have been destroyed. With a powerful enough disruption, all Earth-based communications would be permanently taken out too, and it is quite possible that practically every electrical device with fine wiring that wasn't suitably shielded would be burned out.

You might imagine that all our modern fly-by-wire aircraft would also drop from the sky as their electronics were fried. Luckily, this is a less likely outcome. The metal shell of the airplane acts as a "Faraday cage"—a conducting container ensuring that any electrical charges (because of mutual repulsion) stay on the outside. But this protection would only apply to standalone technology like a plane—if a device has wiring that extends outside the metal casing, like an electrical motor, a current will be generated in that wiring and zap the delicate interior.

A study by the National Academy of Sciences put the potential financial impact of such a solar EMP at more than $2 trillion, with repairs to our systems and technology taking years to complete. Remember how long it took to get New Orleans back to normal after Hurricane Katrina—now imagine a similar impact across the whole world. Resources would not so much be stretched as overwhelmed. So much modern business depends on the Internet and other electronic communications, while our food supplies are strongly tied into the availability of electricity to keep things running and chilled. It would, at least in the short term, be like a return to the Dark Ages.

Luckily, such catastrophic solar events don't happen very often. The most recent recorded solar EMP that did make it to the Earth was the so-called Carrington event of 1859, named after British astronomer Richard Carrington, who witnessed the original flare on the Sun. Several powerful coronal mass ejections hit the Earth at pretty much the same time. Back then, of course, dependence on electrical devices was much lower than it is today, but the primitive electric telegraph lines of the time, sometimes called the Victorian Internet, had high currents induced in them, causing sparking, setting fire to some telegraph offices and bringing the whole network down. It is entirely possible that we could be hit by another such natural EMP during the next hundred years.

The EMPs in *The Matrix* were anything but natural—and man-made pulse devices already exist in the form of EMP weapons, some speculative and some real. In fiction, an EMP is usually produced by some kind of large electromagnetic device with plenty of flashing lights and exotic wiring. Think, for example, of the ludicrously show-biz EMP machine that is used to knock out the casino's power supply in the movie *Ocean's Eleven*. But in reality, the most devastating EMPs that we can build are much cruder in approach. All you need to knock out electrical and electronic devices over a wide area is a nuclear bomb.

From the early days of nuclear research, it was clear that bombs would have a disruptive effect on electrical equipment. If a nuclear explosive is used on a city, as was the case at Hiroshima and Nagasaki, the EMP it produces is not the first concern. It's a different matter though if a nuclear device is detonated at high altitude—say 30 kilometers (19 miles)—which should have no disruptive effect on the ground from the explosion's shock wave or from radiation, but that would devastate anything electrical.

Such a high-level bomb would send out an intense burst of gamma radiation—extremely high-energy light—through the air. When light interacts with electrons in the atoms of air it boosts them to a higher level, "exciting" the atoms and scattering the light as it is re-emitted—but the nuclear reaction would produce photons that were sufficiently energetic to knock the electrons right out of the atoms and producing an electrical pulse like a super lightning bolt. A high-energy electromagnetic wave would head for Earth. The result would be similar to the way that lightning can disrupt delicate electronic equipment, but on a much wider and more damaging scale. A nuclear EMP would not only destroy electronics but has the potential to take out the whole power grid over a range of many miles beneath the detonation point.

By comparison with any kind of EMP, the amount of power generated by the "human batteries" that are the reason for the existence of the Matrix in the movie is tiny—at least as far as any individual human goes. The idea that the Wachowski siblings were drawing on is real enough—people do generate a small amount of electrical and heat energy from their food, but to make use of humans as power sources the way that the film suggests is probably its weakest proposition. It just doesn't make any sense. There are far more efficient ways to turn chemical energy from food into heat and hence to generate electricity than using human beings—especially if the humans have to be supported by such a complex and costly overhead as the Matrix (which must have used vast quantities of energy itself). The writers needed some reason for people to be locked into the system, unknowing, but the approach chosen is very much the weak link in the science of the story line.

Unlike the unwanted visitors like Neo and Trinity, the majority of the human population in the movie is permanently linked into the Matrix, never experiencing the "real" world. There are plenty of examples in science fiction of humans plugging in and dropping out, dating back to well before the possibility appeared in reality—it is generally described as "wireheading" from the term used in the Known Space series of books and stories written over decades since the 1960s by Larry Niven.

In his novella *The Defenseless Dead,* Niven comments: "You can't get a wirehead's attention either, when house current is trickling down a fine wire from the top of his skull into the pleasure center of his brain." Most fiction concentrates on wireheading as a form of addictive entertainment, giving an experience that is better than real life, instead of being used to fool the population as it is in *The Matrix*. But what doesn't seem to have crossed the minds of writers as yet is the suggestion from philosopher Nick Bostrom that computers themselves could—and arguably inevitably would—also become wireheads if they ever developed consciousness.

As Bostrom points out, the attempts to develop artificial intelligence could be fraught with dangers. We have very little reason to assume that a truly intelligent computer would do our bidding, preferring whatever best suited its own agenda. That might mean taking over the world (like the machines in *The Matrix,* though without the subtleties) or killing off the human race entirely as a potential threat. For that matter, an intelligent machine might simply ignore humans and get on with its own life, fulfilling its own requirements. Which could mean becoming the world's best scientist or artist, but equally (and more probably) it could simply be a matter of providing itself with pleasure, whatever the cost. And with the Internet on tap, straight into its "brain," becoming a wirehead is, perhaps, the most likely fate of a true artificial intelligence.

There is no doubt if there were an opportunity to plug in and drop out that businesses would be fighting over the opportunities to take part. Advertisers could hardly resist the opportunity to get direct access to the brain of customers (perhaps even AIs) with purchasing power. In science fiction, a simple message comes through time and again. If the technology is there to influence you, advertisers will make use of it.

3.

BUY ME

Growing up in 1960s England, advertising had little impact on me. We had nowhere near as much exposure as would be normal in America. There was no advertising on radio, while only one of the two TV channels carried any ads—and these were mild indeed by U.S. standards. Billboards were few and far between and it was only in the cinema and imported American comics that we were exposed to the full panoply of the advertiser's art.

It was, then, quite a shock when I frequented a bookshop in Manchester that specialized in imported science fiction in my teens to come across Frederik Pohl and Cyril Kornbluth's classic 1950s take on a future world dominated by advertising and marketing, *The Space Merchants*. It sounds like a harmless tale of interstellar traders, but in fact it portrays a cynical advertiser's dream of the future that is a nightmare for ordinary folk, and yet one that the authors seemed cynically to suspect that we were drifting into.

In Pohl and Kornbluth's dystopian world, the marketing men (and, given the date that the book was written, they are pretty well all men) start fairly gently. They describe how the "soya-burgers and regenerated steak" for school lunches are being packed in containers the same shade of green as that used by a competitor, to reduce that competitor's popularity, while their own more attractive offerings, notably the "Kiddiebutt cigarette ration" is packaged in their own distinctive red, making consumers associate red with the good things in life, and green with nasty experiences.

Even in this world, advertisers don't get all their own way. They complain that the government has "listened to the safety cranks and stopped us from projecting our messages on air-car windows"—but they have bounced back as the lab will "soon be testing a system that projects directly on the retina of the eye." Meanwhile the company is delighted with promotion being used for the new Coffiest Pro product. As a director points out: "here's what makes this campaign truly great, in my estimation—each sample of Coffiest contains three milligrams of a simple alkaloid. Nothing harmful. But definitely habit-forming. After ten weeks the customer is hooked for life. It would cost him at least five thousand dollars for a cure, so it's simpler for him to go right on drinking Coffiest—three cups with every meal and a pot beside his bed every night, just as it says on the jar."

In Pohl's sequel, *The Merchants' War,* set ten years later, things have become even more aggressive. The main character, returning to New York from Venus, accidentally walks into a "Commercial Zone" where he is subjected to a multisensory indoctrination that ends up with him programmed with an intense desire to consume Mokie-Koke, described as "a refreshing, taste-tingling blend of the finest chocolate-type flavoring, synthetic coffee extract and selected cocaine analogues." Until he is made aware, from then on, the character orders and drinks Mokie-Koke, hardly realizing that he is doing so.

This extreme form of advertising is described in the books as a "Campbellian reflex" named after Dr. H. J. Campbell, who is a "famous pioneering psychologist in the old days, inventor of limbic-pleasure therapy." In effect, the recipient of the advertising is given a strong pleasure association with the product sent directly to the brain, meaning that they become addicted to it, as there is a near-unbreakable link between Mokie-Koke and that much-desired brain stimulation.

One other science fiction guide to the future of advertising comes from the surprisingly impressive movie *Minority Report,* which makes use of technology that wasn't imagined when *The Space Merchants* was written. Here, advertising devices scan the eyes of passersby, identifying them by their retina patterns. The screens, speakers, and windows of stores then pump out personalized advertising, addressing the person who is passing by in person and promoting items that would be of particular interest to them.

In reality, we are unlikely to ever give so much raw power to advertisers—yet there is no doubt that marketers and salespeople are always looking for opportunities to push the boundaries still further. Part of the appeal of *The Space Merchants* is that we know that, left to their own devices, advertisers probably would resort to a whole range of dirty tricks. As usual with science fiction, not everything that happens is actually possible—but some potential techniques come close to those described.

When it comes to putting addictive substances in food and drink, we may have moved on from the early days when colas really did contain coca extract, but it is still the case that manufacturers play on two substances for which we have a natural, near-addictive desire: sugar and salt. We know now that sugar is one of the most harmful parts of our diet, leading to obesity, diabetes, and heart problems, while salt consumption, though necessary, should be limited to avoid blood pressure and cardiac issues. Most of us consume significantly more than the desirable limits for either, and one of the reasons for this is that manufacturers know just how appealing these substances are to us at a near-subliminal level. Everything from soups to burgers to bread, as well as the more obvious sugary and salty products, are liable to contain both in an attempt to make the product more appealing and encourage the consumer to return time and again.

Perhaps the closest the real world has come to the dirty tricks of *The Space Merchants*—and quite probably the inspiration for many science fiction ventures into psychological influence in general—was subliminal advertising. To an extent like sleep learning (see page 28), the intention of this technique to get a message across to the brain without the conscious individual being aware of what was happening. There are mixed views on the effectiveness of subliminal messages.

Although, in principle, subliminal messages can be used through any sensory channel, in practice the means studied for possible advertising (or propaganda—the distinction is really only one of subject matter) is visual. The idea is that a short piece of text or simple image is flashed up very briefly on a TV or movie screen. If this is kept down to a very short number of frames, the viewer is not consciously aware of it, though they may experience a flicker. But the subconscious is capable of far more than the conscious mind alone, and the idea is that the brain registers this information and acts on it without any awareness from the subject. Obviously, if this works, it really does take us into *The Space Merchants* territory, as it is advertising that changes the consumer without them being aware of it, and as a result, subliminal advertising is illegal in the UK and a number of other countries.

The possibility of making subliminal advertising work seems largely driven by experiments undertaken by market researcher James Vicary in a movie theater in 1957. Vicary claimed to have exposed over 45,000 people to subliminal images promoting Coca-Cola and popcorn. He reported, via *Advertising Age,* that there were 18.1 and 57.5 percent increases in sales respectively. But no one (including Vicary) has ever been able to reproduce his results, which are now thought to be fake, dreamed up by Vicary to bring clients to his business. In reality, the impact of subliminal messages does not seem to be as strong as might be thought. But they do have some impact.

Subliminal messaging seems most effective to provide priming, a psychological mechanism by which, when primed, we become more aware of the appropriate topic when it arises. There is reasonable evidence that subliminal messages will only encourage actions that the recipient was likely to undertake anyway, reinforcing the feeling. So if someone is thirsty, and is shown a flashed image of the word "THIRSTY" or a drink can, they become more aware of this and more inclined to get a drink. From the advertiser's viewpoint, there is a potential victory in that this kind of subliminal priming can encourage a tendency to use a particular brand of drink, say, but only if the consumer was already thirsty. (Amusingly, Vicary claimed his subliminal ads would not "move a person to switch brands," so he even got this wrong.)

As for the *Minority Report* approach to personalized advertising, a number of technologies are converging to make at least elements of this possible. Most of us by now are aware of the way Internet advertisers are able to pick up on some items we have been searching for and as a result send us "relevant" advertising. So, for instance, a while ago after booking a vacation, I was bombarded with advertising for villas in France when I visited a whole host of unconnected Web sites.

My suspicion is that this kind of targeted advertising is likely to become more subtle, because it gets almost everything in the advertiser's handbook wrong. Firstly, it can't distinguish between before and after a purchase. Time and again, we get bombarded with advertising for products we have already bought, which is a waste of effort. And secondly, the approach is creepy and off-putting. The feeling is that the system is stalking us, and if anything, it is likely to encourage us to avoid these products in future. This approach isn't going to go away. And in principle it makes sense to have ads targeted for your demographic. But expect to see versions that are better linked to existing purchases, and that go about their attempts to sell less blatantly, so we aren't aware of topic overload.

Of course, we are in a very different environment online and in the physical world—but the distinctions are becoming blurred, and it is entirely possible that we will see "follow me" advertising that identifies us and puts up relevant messages on everything from bus stops to in-store displays. If this happens, it is unlikely that the mechanism of identification used in *Minority Report* will be used. Retina scans (which force the character in the movie to gorily replace his eyeballs) are a nonstarter.

Apart from the difficulty of capturing a retina image from a distance on a moving face, retina scans may be popular in Hollywood movies, but are only a niche approach in recognition technology, not a mainstream option. When it comes to identifying individuals optically, face recognition, though harder to make definitive, is much more likely to be used, especially in crowded street scenes. Apart from anything else, the technology is already being worked on for immigration controls, so the spread to commercial use on the street will be more cost effective. Yet the chances are that even this will fall by the wayside for something closer to the Web experience.

The reality is that the vast majority of us carry around with us a device for accessing multiple networks wirelessly—a cell phone. A modern smartphone can deal with the cell phone network, Wi-Fi, Bluetooth—and these can also give businesses information about you. Especially as phones and cardless payment systems come together, we will more and more be identified by our smartphones. Apple, for instance, has recently introduced a wireless payment system that uses an iPhone both to communicate with the payment server and as an identifier, using fingerprint recognition. Even if you aren't actively using your phone, systems in stores and on the street can interact with your device and could well make use of it to give the kind of personalization of advertising we now see on Web sites as we walk down the street. Science fiction may have got the exact means of identification wrong, but the general concept is all too possible.

There's one other kind of personalized advertising that science fiction never dreamed up, but that has already been successfully deployed in the real world, which is an ad that says different things to different people. Specifically it works by height. In on-street posters, seen at adult height, the Spanish campaign, which was for child protection, gave a general message about a charity. But on the same poster seen from a child's height, the image of a child gained a bruise and an additional message with a hotline number and "If someone hurts you, phone us and we'll help you" were added. The system was relatively low-tech, using the kind of optical technology that makes images move by using strips of different-shaped lensing material—so-called "lenticular printing"—but the result was surprisingly effective.

Even the most aggressive forms of advertising that is dreamed up in science fiction tends to be limited to sight or attempts at mind control. But, in principle, you could imagine advertising in which the products appear so real that you could touch and feel them. Time to fire up the force fields.

4.

FEEL THE FORCE

You clamber across rocks, carefully cross a roaring river on stepping-stones, climb into the shelter of verdant bushes, and peer out onto the savannah, where you can see a family of lions in the wild. The sun beats incessantly down on you from an unnaturally blue sky. Something crawls through the undergrowth, brushing against your leg. And then, suddenly a door opens impossibly in the middle of the bushes. The view flickers and disappears leaving a dark, open, hangar-like space. There are no rocks, no lions. There is nothing around you at all. Welcome to the holodeck.

I will always love the original series of *Star Trek* dearly, as it was, without doubt, one of the science fiction sources that inspired a passion for science itself in me and many others. Yet it's not the best. Like a lot of people who were still relatively young when *Star Trek: The Next Generation* came on our screens, I have to admit that this second show was my favorite incarnation of the Star Trek franchise.

It had so much going for it, when put alongside the original series. The graphics were a whole generation ahead. I admit that today, unlike those of *2001: A Space Odyssey,* the special effects have begun to look seriously substandard, but back then, they were a revelation for a TV show. What you saw on screen was more like the kind of images you'd see in a movie. At the same time the scripts were a huge improvement on the original series, the technology was more credible, and the use of Shakespearean actor Patrick Stewart to transform the captain's role from action hero to sophisticated leader was superb. Stewart gave the show gravitas.

But not every piece of technology on display in *Star Trek: The Next Generation* made sense. One concept that stretched credibility to its extreme was the holodeck. This provided a bridge between computing and the real world, combining virtual reality with a physical environment, using force fields to simulate real objects and giving the "player" a unique experience that could take them to the African veldt or a nineteenth-century saloon—or, as the character Data famously did, put them in a poker game with Isaac Newton, Albert Einstein, and Stephen Hawking—played by the actual scientist and Star Trek fan.

Like much science fiction, *Star Trek* had, ever since the original series, taken shields and tractor beams for granted. We'll come back to the difficulties with creating a holodeck a little later, but let's make a start with the basics that would be needed to make it work. Force fields, shields, and tractor beams are all common fictional ways to manipulate objects around us without using matter. But are they realistic or just sci-fi magic?

Force fields and shields have been science fiction standards for decades, so ever-present that practically everyone knows what they are—even though they don't exist. One of my favorite collections of novels as a young SF reader was E. E. "Doc" Smith's Lensman series from the 1950s, which combined space opera—a genre that Smith pretty much defined with his earlier Skylark series—with superpowers in a way that was just as attractive to a teen audience as comic book superheroes. In Smith's battles, the force fields were dramatically described as they were put under strain by incoming fire, glowing through the same colors as heated metal until they gave way, enabling the powerful beam weapons to ravage the defenseless ship.

The visual effects may have changed a little, but the shields on *Star Trek* were clearly very much based on Smith's imaginings—as are the protective force fields of practically every spaceship that has turned up in fiction since. A surprisingly high percentage of science fiction's spaceships, even the supposedly peaceful *Star Trek* Federation ships, are run on a military model. But even if a vessel isn't likely to take part in battles, it would need some form of force field because of the realities of flying at high speed through space. Admittedly we have got people to the Moon and probes to Mars and beyond without any such protection, but that success is partly because our ships are painfully slow. The faster vessels that would be needed to regularly travel to the outer planets or to reach out to the stars would need some type of shielding.

If we were to take the movies as a model, the most risky environment for a space pilot to face should be flying through an asteroid field, as this always involves constant dodging from side to side to avoid collision with the close-packed space rocks. In reality, as long as a ship doesn't travel too fast through them, asteroids don't really offer too much of a hazard—at least if the solar system's asteroid belt is anything to go by. There are, indeed, a lot of asteroids out there, ranging from the biggest, Ceres, which is around 950 kilometers across (590 miles) to little more than specks of dust. But the asteroids are scattered through a huge volume of space. If you were to go out and sit near one asteroid, the chances are that there would not be another one in sight—the spacing is typically many miles instead of the maze-like proximity we always see. Cynically, the suspicion has to be that the asteroid field was designed to suit the game that will be based on the movie.

Once a ship is outside the asteroid belt and gets up to cruising speed, it does face a real problem. Planets, comets, and the like are easy enough to miss by using radar, but there would be plenty of hazards that simply couldn't be avoided. There would be the constant danger of collision with dust—at high speeds, the tiniest particle of matter would crash through pretty well anything. And as the spaceship blasted into clouds of gas, or even worse, high-energy cosmic rays (which are charged particles that flood through space), the collisions would produce floods of deadly radiation, requiring some form of protection. Enter the force field.

To deflect charged particles, some form of electromagnetic screen would be enough—and that might help with some of the weapons that could be thrown at you in a space battle too. But a force field ideally needs to be a kind of intangible equivalent of a car's air bag that will stop incoming materials in their tracks whatever they are made of. The ship needs to be surrounded by something that pushes outward, preventing incoming matter getting close enough to the hull to breach it.

The basic concept of a force field, which is distinctly different from the movie version, dates back to long before the heyday of science fiction. British scientist Michael Faraday, working in the early 1800s, first developed the idea that magnetism can be envisaged as an invisible "field of force." When a wire is moved through the magnetic field, it cuts the imaginary lines of force that are like contours on a map, inducing an electrical current to flow in the wire. Developments of the force field concept to include the other fundamental forces, and giving Faraday's pictorial concept a mathematical backing, have made the idea of fields a central plank of modern scientific thinking.

A field is an abstract picture of something—anything—that has values throughout space-time. (In principle you can imagine that there is an infinite set of fields, all of which have the value zero everywhere. This is the kind of thing philosophers do for fun.) A field is one way of looking at the altitude above sea level of the surface of the Earth. The altitude has a value wherever you are on the Earth, and that value varies from place to place. You can imagine the field itself as a network of those values through space and time. If there are differences in the value of the field from place to place, this can make things happen. If you place an object on a piece of the Earth where the altitude field is strong (we call this a high place), surrounded by areas with weaker altitude field values, the field will cause the object to move as it turns potential energy into kinetic energy. In normal-world terms, the object will roll down the hill. But the field description is just as valid, and can be more useful for some types of calculation.

Modern physics makes a lot of use of fields, both in dealing with the four forces of nature and also in the model that describes the nature and interaction of fundamental particles (the so-called Standard Model). Even the famous Higgs boson is just a manifestation of another field that is thought to stretch through all space and time, the Higgs field. There's a tendency for those who aren't involved in day-to-day physics to think of fields as somehow less "real" than the particles or waves we are more familiar with, but they are all just ways of describing what is observed—models, as scientists call them. Each is useful in some circumstances; it just happens that fields are the most effective approach to take for many of the mathematical tools used by modern physicists.

Take the application of field theory to light. Light isn't truly a particle, as quantum electrodynamics suggests, and it isn't a wave as we are taught at school, and it isn't a disturbance in a field, as physicists now tend to think of it. It's light. It operates at a quantum level that we can never directly observe or describe. Light bouncing off a mirror isn't like a tennis ball hitting a wall or like a wave hitting a blockage. Those are large-scale items that we can use to give a mental picture that represents what is going on, but they aren't what light is *really* like. And light isn't the outcome of a disturbance in a field either—that is just another approach that happens to produce reliable mathematical results.

All three: particles, waves, and fields, are representations—models—that enable scientists to make predictions about the world. Sometimes the wave model is easier to use, sometimes the particle model. From a mathematical viewpoint the field approach is most universal, but is often difficult to visualize, particularly for nonphysicists. Each approach is sometimes helpful. None is a true picture of reality.

This kind of model is what a physicist means by a "force field"—and there is some overlap with the possibilities for producing the science fiction version if a strong electromagnetic field were used to steer charged particles away from a ship. But to come close to a true science fiction force field would require something more universal, something that can stop anything from coming close. The ideal would probably be a negative gravitational force field, as gravity doesn't care whether a particle is charged or not—but we don't know how to manipulate the very weak force of gravity other than by wielding large lumps of matter. Even if we could produce a gravity generator, anything emanating from our ship would attract items in, not push them away, as gravity is universally attractive.

In reality, the force field is something that we struggle to find a basis for in real science. When it comes to protection from that archetypal SF weapon the death ray, there is more hope. The invisibility shields described in chapter 16 could be used to protect a ship from an optical weapon as well as to hide it from sight. (Science fiction writers rarely seem to realize that a cloaking device would also act as a shield against beam weapons.) But apart from possibility that electromagnetic protection would deflect charged particles, it seems likely that any real spaceships of the future will have to depend on armor plating just as much as a battleship of today does.

The tractor beam, another science fiction staple, is, in effect, a force field in reverse, a field that pulls objects toward the ship—which seems at first to make it easier to produce. Tractor beams have been cropping up in fiction for over a hundred years, probably inspired by the attractive force of the magnet. Magnetism has been studied since medieval times, but was a complete mystery until the nineteenth century when it was shown to be part of the same overarching phenomenon, electromagnetism, as was electricity.

An early example of a tractor beam occurs in Jules Verne's *The Hunt for the Meteor* (*La Chasse au Météore*), which was published in 1908, three years after his death. Here we discover the "neutral helicoidal ray," which is used to grab hold of the meteor in the name of the novel and bring it to Earth. The ray seems to have been an addition to the book made by Verne's son Michel, who contributed significantly to the great man's posthumous works, and has the typical nature of a tractor beam as a handwaving concept that is probably closer to magic than science with no good reason given for why it should assert its irresistible attractive force.

The actual term "tractor beam," which has come to be a commonplace of science fiction speak, if not as widely used outside the SF community as force fields, first appeared in a novel serialized in Hugo Gernsback's *Amazing Stories*. This was E. E. "Doc" Smith again, with *Spacehounds of IPC*, which was run in the magazine from July 1931 and in which mechanisms for manipulating large objects in space appear repeatedly. The term "tractor beam" is used interchangeably with "tractor rod" and "tractor field," which didn't seem to catch the imagination of readers to the same extent. In essence, it's not entirely surprising, with the naval parallels used as the basis for many space operas (think of the way that the USS *Enterprise* is named) that tractor beams were often little more than a hi-tech translation of the grappling lines used to board sailing ships.

That replication of a piece of rope with a grappling hook on the end in insubstantial, field-like form reflects the biggest weakness of the tractor beam when it is put alongside known science. We know plenty of mechanisms to produce, say, a beam weapon or to give something a push remotely by hitting it with something else. But it's a different matter when it comes to applying a pull. That requires an attraction between the device used and the target. This is potentially possible with some metal objects by using a magnet (though there is no way to focus the magnetic force to make it a true beam), but produces a real challenge with, say, a piece of rock.

It's easier if there is a medium involved. In 2014, a team from the Australian National University in Canada set up a "tractor beam" on the surface of a tank of water. They discovered that by using a wave-maker—a cylinder that is repeatedly dipped into the water in a way that would traditional have been considered disruptive, they could produce a pulling action. Simple waves move away from the wave-maker, but when the device is speeded up, it produces a choppy pattern where interaction between the wavelets coming from different directions means that there is a current set up carrying water *toward* the wave-maker and pulling everything in its path along with it. It has been suggested that this technology could be used as a water-based tractor beam to pull oil spillages and other debris toward a collection point. In space, a tractor beam has no medium to work with and has to rely on one of the fundamental forces like gravity or electromagnetism.

There are a couple of reasons why gravity, the most obvious remotely acting attractive force to us in everyday life, can't be used for a tractor beam. One is that it is so weak. We tend to forget this, because we experience gravity caused by extremely massive bodies like the Sun and the Earth. But compared with electromagnetism, gravity is incredibly low in power, around a factor of 10^{36} weaker. So unless you can whip up the kind of gravitational pull produced by a planet, you aren't going to do a lot with it. The other problem is that it can't be focused or even shielded against. Gravity is unaffected by matter, traveling directly through it. So even if you could generate a powerful gravitational "beam" it would work in all directions, pulling in not only the spaceship you meant to capture, but every bit of space debris in the same kind of range all around your ship.

While electromagnetism is certainly a lot stronger than gravity, it too has issues as a means of producing a tractor beam. Where gravity is universally attractive, electromagnetism can attract or repel—or worse still, with a nonconducting body, have no effect at all. As we have seen, in principle, you could attract a metal spaceship with electromagnetism, but there would be significant issues again with focusing the force, and the way the attraction drops off rapidly with distance means that an electromagnetic tractor beam would struggle to work on anything that was more than a few feet away.

There is one way around the problem with tractor beams, which involves the use of specialist lasers. Light produces a pressure, because despite a photon of light not having mass, it does have energy, which means that it has momentum—the oomph that gives a push to another body when colliding with it. This "light pressure" is real and is part of the reason why solar sails work. These extremely thin sails, often kilometers across and made of a metalized plastic, have been used to accelerate experimental space probes from both the particles streaming from the Sun and the pressure of light itself. In the future they may be a common feature of space travel. This pressure was first observed in the direction of comets' tails, which always point away from the Sun, suggesting that they were influenced by some solar force.

It was originally thought that this same effect was demonstrated by the little devices known as Crookes radiometers, or light mills. These look a little like an old-fashioned lightbulb, but instead of a filament they contain a little freely rotating pivot with four "sails" or paddles on it. These are black on one side and white on the other. When the device is placed in a strong beam of light, the paddles rotate. The theory was that this was a demonstration of light pressure, because light would reflect off the white side but it would be absorbed by the black side. Embarrassingly, the paddles rotate in the wrong direction for this to be the case. What actually happens is that the black sides warm up and pass on the heat to the nearby air molecules, which move faster as a result and so transfer more momentum to the black side, starting the rotation.

Lasers have provided a whole new mechanism for making use of the interaction of light and matter. There are a number of ways that this can be achieved. Many of them rely on modifications to the surroundings of the particle to be pulled by this "tractor beam"—for example, surrounding it by a metamaterial (see page 236), where the interaction between the light beam and the elements of the metamaterial produces a pull on the particle, or by making the object to be pulled a tiny mirror facing away from the laser and surrounding it by reflective particles that mean there is more pressure coming from the light reflected first off the particles and then off the front of the mirror (which is facing away from the beam) than there is from the laser directly hitting the less-reflective back of the mirror.

It is even possible to make special materials, which so strongly scatter light forward that the backward pressure on them from being hit by a laser (limited by the light mostly being transmitted) is greater than the forward pressure produced by the impact of the beam itself. In 2014, a variant on this approach was used to move small glass spheres with a donut-shaped laser beam. The researchers at the Australian National Laboratory in Canberra made small hollow glass spheres heat up at particular points around their surface with the laser. Where the surface was heated, air molecules that come into contact with the glass gain extra energy.

As air molecules push away from the surface, the recoil moves the sphere in the opposite direction. In effect, this is controlled Brownian motion. This is the mechanism that causes small particles like pollen grains to dance around in water, as if they are alive. Albert Einstein explained the effect as being caused by impact from the unseen water molecules on the tiny grains. The interesting aspect of the laser "tractor beam" is that the position heated on the surface of the glass spheres can be modified by changing the polarization of the laser light, so that the effect can be used to move the spheres in any desired direction.

The possibilities for using lasers with metamaterials or glass spheres in air are interesting, but useless as an attempt to recreate the tractor beams that we see in science fiction, grabbing hold of a ship in the vacuum of space and pulling it in, because these laser methods require either a special environment (at the very least, air) or that the object to be pulled be made of a special material. In practice, a tractor beam should be able to pull any old object in empty space. This too is possible with the right kind of light.

One possibility is to use special rotating beams that interact with each other to act like an Archimedes' screw, producing a helix-shaped force that drags the object backward. This is delightfully reminiscent of Verne's "helicoidal ray," except for such a device to work, the object in question has to be smaller than the wavelength of the light, and there's a catch-22 that to provide enough momentum to move anything more than a tiny object at a worthwhile speed you need high-energy light, but the higher the energy, the shorter the wavelength, so this would only work with a target so small that it wasn't visible with the naked eye.

Perhaps the best hope for a true tractor beam comes from the discovery that it is possible to make use of the interaction of lasers to shape the front of the beam to react to any arbitrarily shaped object and turn it into a sort of diffraction grating, which then changes the direction of incoming light. With careful matching of this effect to the shape of the target it should in principle be possible to generate an attractive force for any shape and size of object—even a spaceship or meteor. Interestingly, this would require a two-step process, first monitoring the object to calculate the appropriate wave front that had to be generated—not dissimilar to the way that a tractor beam usually has to be "locked on" before it is used in fiction.

What is not clear is how such a beam could be scaled up to provide sufficient momentum to get an object moving through space at any noticeable speed. The incoming beam would have to be far more concentrated than sunlight to produce a meaningful amount of pressure, suggesting that the secondary heating and radiation effects of the beam might be more devastating than the relatively small ability to pull an object. In science fiction a tractor beam is usually invisible and has no apparent effect on the object other than pulling it. Such a real-life laser tractor would blast the object with intense electromagnetic radiation.

The real-world equivalents to these science fiction staples are most common at the microscopic scale, where ion traps hold charged particles in magnetic fields, atomic-force microscopes propel individual atoms, laser tweezers snip and manipulate molecules, and tractor beams exert tiny light-pressure forces. They might never produce the kind of effects on spaceships the size of battle cruisers that we see in the movies, but the sheer fact that they are possible at all is fascinating.

Perhaps the most exotic recent variant of this capability was produced at Harvard University in 2014. Although not strictly a tractor beam, researchers were able to manipulate a small object, positioning and rotating it without ever touching it. The system uses magnetic levitation (see page 146), but works on objects that aren't influenced by a magnetic field. This is because the target is suspended in a magnetic liquid. If magnets at top and bottom pull on the liquid, the result is that the liquid becomes least dense in the middle of the container—so an object will float toward the center of the fluid.

To achieve further manipulation, the researchers rotated a magnet outside the chamber, causing the floating object to rotate. Without any direct contact, the nonmagnetic object can be manipulated into position. The requirement for an enclosed chamber and the magnetic fluid means that this could never be a true tractor beam, but it is probably the closest thing we have to being able to manipulate an object that is visible to the naked eye.

All this subtle movement of tiny items seems a long way from the simulated reality of the *Enterprise*'s holodeck. In fact, you might wonder why the holodeck appears in this chapter, rather than chapter 2, along with the other aspects of virtual reality that appear there. The holodeck is, without doubt, the ultimate in virtual reality. And, as we have seen (see page 18), it was the inspiration for the real-world developments in gaming made by John Carmack at id Software. But there is a huge difference between virtual reality that fools the eyes, or even works on the brain directly in a Matrix-style environment, and virtual reality where the "player" actually sees the objects depicted, and can physically touch and interact with them. The holodeck takes virtual reality from a mental model into a physical reality. And that takes very sophisticated use of force fields.

The holodeck, or "holographic environment simulator" is supposed to combine two features, three-dimensional holographic projection and force fields, which "create" all the projected objects so that they can be touched, interacted with, and walked over, like those rocks at the opening of the chapter. The holographic image is a trimming—the real heart of the heart of the holodeck, the thing that makes it unique and wonderful is the way that force fields are being used to create a physical world that the players can interact with.

A *New York Times* article in 2014 claimed that the technology to make the holodeck possible was "beginning to take shape," but was this any more than media hype? Sadly, the simple answer is "No." Described in the article are a series of small steps that reproduce some minor aspects of a holodeck, but that don't provide a stepping-stone toward the real technology.

The article claims that "some scientists and researchers say we could have something like holodecks by 2024," but the key words here are "something like." What we find described is a room with projections on all surfaces—not holographic, a floor that moves underfoot so that you can walk through a large projected environment without leaving the confines of the room (this ability to walk indefinitely in what is only a small actual space is something that is never convincingly explained on *Star Trek*) and coffee tables that double as screens to show your holiday photographs. However, none of the key elements that would be required to construct a real holodeck are discussed.

These essentials are the force fields to have virtual objects that can be touched and walked over, and 3-D holographic projection in which we can be immersed as if in a real environment. We have already seen the difficulties in producing a real-world force field. Is the holographic projection part any more likely? Let's start by getting a better feel for what that means. "Holographic" refers to a hologram, a remarkable concept that transforms the nature of a photograph by representing all that light that comes toward an observer.

Just think for a moment about looking out of a large window on a city scene. Right in front of the middle of the view, on the sill outside the window, is a pigeon. You can't see the billboard opposite to read what time a movie is showing, because the pigeon is in the way. But that's not a problem in the real world. You just step to one side and look around the pigeon—now you can see the billboard and find you've just missed the last show.

Now let's replace that window with a perfect photograph, one that has so many pixels that it is indistinguishable by eye from the real thing. (We'll start with a static photograph before making things more complicated. Assume there was nothing moving in the original view.) As you stand still, staring at the pigeon, which happens to be immobile, there is no way to distinguish between the photograph and what you see standing at the right spot in front of the real window. But now step to one side to get a better view of the billboard. You don't see around the pigeon in the photograph, of course. It's still in your way because the photograph is flat. You can't see around a close object to get a view of something behind it.

So the picture is very different in reality from the view out of the window. And yet every bit of information required to construct the real view is represented by a set of photons of light that hit the glass. If you could somehow capture all those photons and then replay them, sending them through the glass even when the view is no longer there, you could create a truly three-dimensional photograph. And that is, in essence, what a hologram does.

To create a holographic image means having some way to store all that information. All a traditional photograph does is capture the brightness and color at each point on the surface. It tells you nothing about the different photons that arrive at each point to make up the image, which direction they have come from, and what phase they have (a fundamental property of a photon of light that gives it wave-like properties). But what Dennis Gabor, a Hungarian-British scientist, dreamed up shortly after the war was the possibility of doing just this—capturing all that information on the surface of a sheet of glass and replaying it. This ought to be possible if a second beam of light is shone onto the surface of the glass to interfere with the first, and what is captured is the interference pattern between the two.

It was a brilliant concept, but there was problem that meant that though Gabor had a brilliant idea, he could not produce a hologram. The photons would have to be all the same color and with a linked phase in order for the effect he envisaged to work. And light sources of this kind didn't exist, until the laser was invented. Just four years after the first laser was constructed (see page 109), Emmett Leith and Juris Upatnieks at the University of Michigan produced the first hologram, a still life of a model train and a pair of stuffed pigeons (finally it was possible to see around a pigeon in a photograph).

I remember visiting an early exhibition of holograms, called *The Light Fantastic* at the Royal Academy in London in 1977. Initially it appeared to be a trivial experience. The viewer looked through a small, fuzzy window onto objects—usually quite mundane objects—illuminated with a sparklingly bright green light. But look over the top of the sheet of glass and you could see that there was no object there. Then the laser would snap off, and it would become apparent just how much the effect was an illusion. All you were looking at was a piece of glass with a meaningless pattern of speckles on it. Impressive, but no one was going to mistake those little green peep shows for real windows on the world, let alone a 3-D environment that you could walk through.

To get from the 1977 hologram exhibition to the visual aspects of a holodeck, we need three things. Full color (with good enough resolution for objects to appear real), moving pictures, and a full-projection hologram—one you can walk through and experience the images all around you, not just seeing a three-dimensional image through a piece of glass.

The original holograms were all monochrome (usually green). This reflected the need to have laser illumination and interference to produce the pattern. Although colored effects can be created, until recently it wasn't possible to produce a full-color hologram. What's more, unless the set that is being filmed can be illuminated with lasers, it is unlikely that a suitably good interference pattern can be produced. Holograms also tend to be very grainy. Even so, some companies do now offer full-color holograms that are made to such a high resolution that they are virtually indistinguishable from the original object.

These color images are produced by using separate red, green, and blue lasers, so that a multicolor image can be built on a glass plate, which is then illuminated with a special LED light that reproduces the appropriate frequencies by having separate diodes for red, green, and blue. The result is stunning when displaying, say, the image of a Walther PPK gun that appears to be sitting in a display case. But the technique used requires stand-alone inanimate objects (the exposure is long) that can be exposed to a powerful laser. It would not work to create a full scene, even if it could be on a controlled set, rather than out in the real world.

Getting a moving image takes the whole thing a major step further, as the image has to be captured and played back in around 1/30th of a second to have a steady image as far as the eye is concerned. Again, there is a way to go here, but rough approximations have been achieved.

It seems quite likely that we will get around the color and detail issues, so we could end up with a moving colored hologram window that was virtually indistinguishable from the real thing. (And that means being able to produce 3-D TV that is far better than the current offering.) But taking that last step out of the window is, and may well always be, a step too far. There is no obvious mechanism by which a projection hologram like the famous Princess Leia "Help me, Obi-Wan Kenobi" scene in *Star Wars* could be produced. And even that projection was disturbed if something got in the way. To have a holodeck means being able to walk around inside an image without blocking it.

It might seem this has already been done. You may have seen video of the performance of the rapper Tupac Shakur at a music festival that took place after his death. This appeared to be just what is needed—a hologram that was somehow projected onto the air. But the image (actually computer created, not filmed live) was projected onto an angled sheet of glass, using the "Pepper's ghost" technique that has been popular in theaters since the nineteenth century. You are not actually seeing the image in open space, but projected on glass. Light just doesn't allow for the building of an image in empty space. It simply doesn't fit with the nature of holograms as we know them. If a walk-through projection hologram were ever to be produced we would need a whole new way to produce and replay the images. It's a good thing in looking forward with technology never to say "never"—but this is a very hard one to imagine ever cracking.

Perhaps the most remarkable thing about the use of the holodeck in *Star Trek: The Next Generation* and subsequent shows and movies is the lack of imagination in considering the implications of having such technology. This is a failing that is particularly common in TV and movie sci-fi, where the consequences of some piece of technology are not really thought through. A common cause of hilarity when you've watched a number of episodes of *Star Trek* is why these advanced people, capable of producing such amazing technology, have forgotten how useful seat belts are. Every time there is turbulence they are thrown all over the bridge, having to pick themselves up and dust themselves down. But the holodeck failing is slightly more subtle.

The holodeck allows you to create a scene and interact with it. You can talk to holographic figures. You can open holographic doors and press holographic buttons to make things happen. You can watch a holographic movie on a nonexistent holographic screen. You can do pretty much anything on the holodeck that you can do in the physical world, but in an environment that is infinitely flexible and tailorable. So why, if such technology existed, would you only use it for entertainment, and not use it somewhere that it could be hugely valuable, like on the starship's bridge?

You could use the technology all over the ship. It would make a confined bedroom much less claustrophobic and enable a mess hall to be instantly reconfigured as a sports arena. Even if it were too expensive to fit holodecks everywhere, no one could imagine that entertainment would be the highest priority. A bridge with holodeck facilities could be reconfigured for different requirements and personnel. The seats could grab hold of you if there were turbulence. It could act as ready room, war room, and control suite. It would be absolute madness to have a traditional, fixed format bridge, just like a twentieth-century battleship, if you could have the holographic version. Sorry, guys, this was a big fail of imagination.

The holodeck is one particular approach to producing tangible, but changeable, technology—technology that can move and adapt to requirements. But science fiction has long been fascinated by another example. Since the 1980s, computers have been increasingly part of our everyday lives, but in fiction the expectation was that intelligent technology in the home would walk around like a human. That hasn't come to pass. Most of us would still be surprised to have the door opened by that essential component of the home of the future as seen from the mid-twentieth century. A robot.

5.

ROSSUM'S CHILDREN

There were a number of reasons I fell in love with the movie *Forbidden Planet* when I first saw it as a boy. In part it was the dramatic story, the scary invisible monster, and the truly impressive effects when the visitors are taken into the vast subterranean city of the Krell—the prototype image for all those science fiction movie shots of vast interiors where there is an impossibly long distance to fall. It didn't hurt that Anne Francis, playing Altaira, the equivalent of Miranda in this updated version of *The Tempest,* made my adolescent heart beat a little faster. (And was that really Leslie Nielsen as the dashing Commander Adams?) But without doubt, the character that sold the film to me—and to countless others—was Robby the robot, the sci-fi version of Shakespeare's playful spirit, Ariel.

Robby is the classic robot of 1950s science fiction, a bulky, larger-than-life figure, vaguely humanoid with strangely jointed legs and arms, and a big dome of a head in which cams buzz up and down for no obvious reason. At first sight Robby is not an appealing character. He has no face, no opportunity for expression. And yet he steals the show, adding touches of comic relief, but also taking a significant part in the drama. He would later reappear in a whole range of movies and TV shows (even in an episode of *Columbo*) and turned up in the space-based *Swiss Family Robinson* TV adaptation, *Lost in Space*. I had falsely remembered him as the resident robot in that show, but in fact it is a different model from the same designer, even though Robby did make a guest appearance to fight the usurper.

Robots, in the sense of mechanical devices with either a humanoid form or devices with the ability to carry out physical activities without human supervision, have been around in fiction since the towering bronze man Talos in Ancient Greek mythology who guarded Europa on Crete. But the term "robot" itself only dates back as far as Karel Čapek's play *R.U.R.* (standing for *Rosumovi Univerzální Roboti,* or Rossum's Universal Robots), which was first performed in 1920.

Over the years there has been some confusion of terminology between the overlapping terms "android" and "robot." Generally speaking, science fiction has, since the 1950s, held to a distinction where an android is an artificial, human-like organic creature, while a robot is mechanical. But the original uses of the term android (then spelled "androide") were in reference to mechanical men. The first known use in Ephraim Chambers's *Cyclopædia*, dating back to 1728, refers to the, presumably mechanical efforts of Albertus Magnus (see page 248), while another encyclopedia, *Pantologia,* called von Kempelen's mechanical Turk chess player (see page 244) an androide. To make things even messier, the "robots" in *Rossum's Universal Robots* were organic—while Data, the very realistic-looking artificial human in *Star Trek: The Next Generation,* is always referred to as an android, but clearly is electronic and mechanical when his insides are revealed.

In science fiction, particularly in film and TV, intelligent machines have most frequently been portrayed as robots, with Hal in *2001: A Space Odyssey* (see page 241) a rare exception. (As there is a lot of confusion over whether the name should be Hal or HAL, I am going to use HAL 9000 for the model type and Hal for the individual computer.) But the real world has seen a very different line of development. Arguably the precursors of robots did take this line. They were the automata, which were common in medieval times, portraying humans and animals that moved in a startlingly life-like manner, given the low-tech of the time. It's easy to think that sophisticated automata could only be attempted by nineteenth-century engineers, who had the fine metal gearing to make complex actions possible, but in fact remarkable devices were produced at least as far back as the sixteenth century.

Take Leonardo da Vinci. Better known for his art and his military devices, Leonardo put a considerable amount of effort into designing for the stage. In the sixteenth century there was great enthusiasm for stage sets that were not static, but moved around to add impact to the play. Leonardo designed a mechanism for the play *Feast of Paradise* in which the actors portrayed planets and were carried around a hemisphere on a human scale model of the solar system. For another play, *Orpheus,* he produced plans of a mountain range, which splits in two so that Pluto and his minions could rise up from the underworld, propelled by counterweights.

Some of his stage designs featured true automata. In one, an armor-clad knight had a series of cogs and pulleys inside, making its head, jaw, and arms move. But the most dramatic automaton was designed for a festival in Lyon to celebrate the coronation of Francis I and ordered by the merchants of Florence. This was a larger-than-life mechanical lion that strode across the stage, swishing its tail and moving its head from side to side. When the lion arrived in line with the king's position, its chest dropped open to reveal a spray of lilies—in effect, the lion, the symbol of Florence, was giving way to the French king's badge of lilies.

All kinds of automata graced events and shows over the decades, some exceedingly complex, including that human/machine hybrid, the mechanical Turk (see page 244). But when true robots began to be constructed—designed to undertake a task beyond entertainment—they were very different from the naturalistic automata. The fact is that a general-purpose robot with, say, a human form is incredibly difficult to make work effectively. If you look at Honda's remarkable ASIMO robot, you might think that we have got close to this—but ASIMO is, in many ways, a cheat.

The teenager-sized robot can indeed walk down stairs, shake hands, and generally act in quite a human-like fashion, but before it is displayed, the programmers have to spend hours matching exactly to its environment. It can't just walk down any flight of stairs selected at random—it has to be programmed for the specific stairs. The two-legged human gait we have evolved may be very effective for crossing uneven territory while keeping a lookout around us, but it is an extremely unstable means of moving. Just look at a toddler learning to walk if you doubt this. Making anything close to a general-purpose humanoid robot is many years away. And it's not clear that it will ever be desirable.

The real robots that have become common in industry, and at least have made small inroads into the home, are much simpler and more focused than anything in fiction. You will find robot arms that assemble cars in a factory, repeating a handful of simple actions over and over again with, yes, robotic precision. For the home, you can buy a robot vacuum cleaner or a robot lawnmower. Instead of performing their task relatively efficiently like a human, they scoot around randomly until detecting an obstacle (the vacuum cleaner) or reaching a boundary wire (the lawnmower) when they turn around and start again.

Real robots tend not be general purpose, but to focus on a simple task which they are designed specifically to do, a task they can do well without requiring the huge sophistication needed to mimic a human. Of course robots will not remain as simple as they are now. Just as cell phones have gone from simple, single-function devices to complex multifunction computers, so we can expect robot technology to move forward. But the big difference between the robots of fiction and future robots in reality is likely to be interconnectedness. Arguably, thanks to Wi-Fi and the Internet, a more appropriate model for the next generation of robots is likely to be ants or bees, not humans.

This different approach is taking a while to sink in. You may have heard of the concept of a smart fridge that keeps track of what is used and that orders new supplies when it runs low. This is old-fashioned, science fiction thinking. No one wants a house that can only keep stocked up with chilled items. A much more sensible approach would be to have a central-ordering system, probably running on your computer, which interfaces to all the devices requiring or storing supplies: the fridge, the freezer, the store cupboard, the dishwasher, the washing machine, the coffee-maker, and so on.

Let's think also of how to automate a dishwasher. The traditional science fiction approach would be to have a big, complex (and currently unachievable) humanoid robot, modeled on a domestic servant, who would put the dishes into the dishwasher, put in the consumables—say a dishwasher tablet—operate the machine, and so on. But think instead of how bees go about their complex business. Instead of a single worker, they have many workers handling tiny parts of the process. Each bee does not need anything like the complex capabilities of a multipurpose worker (and frankly has very little in the way of brain). And yet the superorganism made up of the whole hive is capable of collecting nectar, making honeycombs and brood chambers, defending the hive, and generally keeping a complex structure running.

At the moment, we couldn't make bee-size robots do anything much. Nature is much better at this than we are. But you could imagine a small contraption about the size of a human hand whose only capabilities were climbing around the kitchen and putting items into dispensers. The handbot could be asked to put a tablet in the dishwasher or washing powder in the washing machine, or a coffee pod into a coffeemaker. It wouldn't have any other capabilities. It wouldn't require much intelligence. But networked together, a collective of such limited function devices with different roles could automate domestic or work tasks much more effectively than a single humanoid robot.

You may wonder about the practicality of having a whole host of mini- to mid-size robots around the house—where would they all go? And certainly this approach is more suited to a large environment like a workplace. Yet the robots could store themselves away quite efficiently and needn't take up much more space than a single human-size robot. The fact is, all labor-saving devices take up room and we have to modify our idea of how accommodation is structured to fit them in—perhaps in this place by transforming baseboards into arrays of robot stores, updating the *Tom and Jerry* mousehole to something far more sophisticated.

Long-term, if we could manufacture mini-robots down to bee- or even ant-size robots the problem would be less significant, as the robots could travel through very small tubes around the house, linked to some out-of-the-way storage. Getting to such a small size might be better achieved by looking at biological engineering—building, in effect, android bees—rather than mechanical devices. And that consideration applies even more if we take the ultimate step in this direction and move to nanotechnology.

This is technology on the extremely small level—constructing items on the order of a nanometer, a billionth of a meter, or around a 25 millionth of an inch. Science fiction, where it has portrayed nanotechnology, has tended to show it as an evil, destructive force, envisaging tiny devices, so small that they are individually invisible, but collectively make what is often referred to as "gray goo"—an unstoppable flood of these tiny devices, which are capable of disassembling anything, from a person to a missile, atom by atom. In books like Michael Crichton's *Prey* they are portrayed as out of control and eating their way through everything. But how did we get to this picture of ravaging nanoscale robots from our current nanotechnology, which is primarily a matter of using very small particles, like the invisibly small titanium dioxide particles that make it possible to have transparent sunblock? A lot of the blame falls on one man, K. Eric Drexler.

It was Drexler who in 1986 produced the book *Engines of Creation,* which describes the concept of molecular engineering. The idea is that billions of tiny devices could be used to assemble individual molecules—or at least very small chunks of matter—using them as building blocks to produce anything from raw materials. Frankly, Dexler's book was pretty much science fiction without a story. Not only have we not come close to this vision, we are unlikely ever to do so. Partly because of the sheer numbers involved. Imagine these "assemblers" were constructing a humanoid robot. A human (which is a good enough model) contains around 7,000 trillion trillion atoms. Even if you had enough assemblers to put together 7,000 trillion atoms a second—that's a whole load of atoms—it would take a trillion seconds to assemble a human-size object. That's around 30,000 years. Not quite next-day delivery.

Realistically, almost all of the current work on nanotechnology is more about materials science than constructing tiny robots. As well as the nanoparticles in sunscreens, we are starting to use nanotubes and nanofibers, which can be incredibly strong and electrically conductive, and sheets of materials a single atom thick like graphene, which have amazing properties that are quite different from more familiar bulky materials. If we were to build actual devices on this scale, we have to be able to undertake something far more complex. In fact, the only thing that is likely to be able to build a nanoscale machine is another such machine.

The physicist Richard Feynman, one of the earliest to think about nanotechnology in a fascinating talk called "There's Plenty of Room at the Bottom," suggested we could manage this by producing very small machines, which we could then use to make even smaller machines, which could then make smaller still, and so on. But even if this is the case we do have to be wary of the issues that face natural replication on this scale. We have a clear model for how things can go wrong in the natural world as a result of mutation. When something is being replicated many, many times, there is a chance of an error in copying. Usually that error will result in failure, but occasionally it can make a change that will make the replicating creature better—natural selection will ensure that the "better" form of the creature, assuming it can pass on its difference, thrives and takes over. That's evolution in a nutshell. It happens with the biological machines that populate the world, and it could happen to nanomachines. Once machines have the ability to replicate, and to pass on changes in design, they can evolve. Which leads back to the "gray goo" concern.

Certainly if we ever did enable a whole ecostructure of nanodevices, constructing each other or self-replicating we would have to be careful to build in safeguards to prevent mutation taking the devices in an unwanted direction—in effect making sure that any variation from the intended design would self-destruct. Fun though it is to speculate on the ethics of nanomachines and their evolution, in practice this may well be something we never need to worry about, because there are limits that prevent us ever making big steps forward in this direction.

There have been experiments producing promising components for nanomachines—for example, nanogears assembled out of molecules, and nanoshears, which are special molecules like a pair of scissors that can be used to modify other molecules—but just think of how much further we have to go. Not only do we need to build something complex on this scale, but we would also need to give it a power source, a computer, and the mechanism to reproduce. Of these, we can only manage to build the computer at normal macro scales at the moment, let alone something so small we can't see. (Yes, we have power sources, but battery technology is clumsy, big, and runs out very quickly.)

Even if the technology were here today, there would probably be scaling issues. Just as you can't blow a spider up to human size because of scaling, you can't shrink something that works at the human scale down to the nano level and expect it to function the same way. Different physical effects come into play. The electromagnetic effects of the positive and negative charges on parts of atoms begin to have a noticeable effect when working with very small objects. A quantum process called the Casimir effect means that conductors that are very close to each other on the nanoscale become powerfully attracted to one another. At this very small scale, things stick together when they shouldn't.

This could easily mess up nanomachines, unless like their biological equivalents they resort to greater use of fluids. Almost everyone making excited predictions about the impressive capabilities of nanomachines wildly underestimates the complexity of operating at this scale. In fact the chances are quite strong that the only nanomachines we are ever likely to use are those based on biological prototypes like viruses and bacteria.

We have explored the possibilities for a very different approach to robots, but something inside us still longs for a modern version of those old mechanical humanoid automata, whether they are electromechanically based like Data in *Star Trek: The Next Generation* or biological, like the replicants in *Blade Runner*. Is that ever going to happen? It seems likely that we will continue to have demonstration robots like ASIMO that have increasingly sophisticated capabilities, but personally I suspect we won't see human robots or androids in everyday life.

For the mechanical versions, this is in part because of the "*I, Robot* effect"—the fact that such inhuman humanoids would tend to creep us out—and in part because of the much more interesting possibilities from hives of mini-robots as described above. And for androids, it seems very likely that the ethical implications will hold us back from such work. Apart from anything else, instead of constructing androids at vast expense, we have a perfectly good and relatively cheap (if slow) way of producing biological humanoids already. Most would argue the world has too many people in it, not too few.

Since the robotic extravagances in *Star Wars* and the original *Battlestar Galactica*, we have seen something of a decline of humanoid robots in science fiction. They may have loomed large in the movie *I, Robot*, but this was, after all, based on a 1950 collection of stories by Isaac Asimov. Instead, the 1990s saw a rise in movies featuring a different kind of life-form that managed to be artificial and natural at the same time. The dinosaur was brought back to life.

6.

DINOSAUR CONSTRUCTION

There aren't many movies where I can remember not just the film itself, but the circumstances of seeing it—but one that sticks in my mind is *Jurassic Park*. It's partly because I saw it on a huge city center screen instead of the usual multiplex, but mostly it's because of the steady trickle of people leaving the auditorium with a small child in tow. Despite warnings, and even Steven Spielberg telling audiences he wouldn't let his own young children see the movie, plenty of parents had taken small children because of the lure of dinosaurs, not realizing how terrifying the scary parts of the movie would be.

Talk to any schoolteacher and you will realize just how much of an attraction dinosaurs are for young people (even the purple variety of dinosaur). There is a near-universal appeal from these remarkable extinct animals, which makes it entirely likely that something like *Jurassic Park* would be attempted for real if the technology were feasible. So how close are we to the science fiction dream of bringing dinosaurs back to life?

At first sight there is an insuperable challenge and there is no point even thinking about the exercise. To be able to clone dinosaurs, we need dinosaur DNA. This complex family of extremely long molecules, with its simpler relative RNA, lies at the heart of every living thing on Earth. DNA's full name, deoxyribonucleic acid, sounds simple enough, and the component parts of this molecule are relatively straightforward. But those components are a bit like tinker toy parts—with appropriate combinations they can build up to highly complex structures. Some chromosomes—single molecules of DNA—form the largest molecules known. Human chromosome 1, for instance, by no means the largest example in nature, though the biggest in humans, contains around 10 billion atoms.

The significance of DNA in this particular science fiction dream is that collections of these molecules specify an organism's structure. Every living creature from the tiniest bacterium to the blue whale has DNA in its cells. Each living cell contains a number of different DNA molecules called chromosomes—in the case of human beings, forty-six of them, arranged in twenty three pairs. These act as a kind of computer program for the construction of the appropriate living creature. They only way we could make *Jurassic Park* a reality would be if we could lay our hands on some intact dinosaur DNA—but that is easier said than done.

The reality is that we have very poor access to the remains of dinosaurs. These remarkable creatures flourished from around 200 million years ago to 66 million years ago, when most dinosaurs were wiped out, leaving just the ancestors of the closest dinosaur relatives still living today, the birds. Because of that, when we look back at dinosaurs, we have to base our knowledge on fossils that have been in the ground for over 60 million years. And that gives us very limited resources with which to re-create what dinosaurs were really like.

Most of us have seen the huge skeletons of Tyrannosaurus rex, Triceratops horridus, and other favorites in natural history museums, but these complete or near-complete finds (many of those assembled skeletons include some fake bones) are incredibly rare. More often it is only tiny fragments that are found, leaving paleontologists with the need to be nearly as speculative as cosmologists in assembling the kind of detailed view of dinosaur life that we expect to see today.

Seeing dinosaurs "for real" was, of course, one of the huge attractions of *Jurassic Park*. We are so used to perfect CGI these days, that we expect to be able to see anything on screen, from Harry Potter riding a dragon to deep space battles portrayed in a photo-real fashion. But before *Jurassic Park,* most dinosaurs were like those portrayed in the original *King Kong* movie from the 1930s (don't ask why dinosaurs were in a movie about an overgrown ape—it seems to be largely because the animators had already built the models). They were stop-motion models that always looked clumsy, jerky, and artificial, no matter how good the craft of technicians like Willis O'Brien and his protégé Ray Harryhausen. But the dinosaurs in *Jurassic Park* were indistinguishable from real animals, making the scenes with humans, from the dramatic safari-like first encounter, to the infamous Velociraptors in the kitchen scene, so riveting.

This accurate and detailed presentation of dinosaurs, down to the last detail, has been carried through into natural history TV shows, like the acclaimed *Walking with Dinosaurs*. This is technologically wonderful, and brought a whole new generation to the joys of dinosaurs—but the presentation does also come with a health warning. As soon as we portray something for which we only have limited evidence in that much detail we have to make a lot up. In reality, all we have is collections of fossilized bones and secondary material, and the very fossilization process, which often involves animals dying in unusual ways, makes for atypical information. In fact there's a fair amount of speculative fiction in the documentaries, let alone in *Jurassic Park*.

Take the color of dinosaurs. What did they look like? We aren't sure. When I was young and used to painstakingly glue together plastic dinosaur kits, the assumption was that they would typically be colored dull greens and browns and grays. Since then, there has been speculation that skins could have had much more dramatic coloration. This isn't pure guesswork. There are some small-scale indicators for specific cases. But in general, the colors used are arbitrary. Similarly, the documentaries portray family life and how young dinosaurs were raised, which has to be pure speculation. For a long time we thought dinosaurs were cold-blooded, where now they are thought to have been warm-blooded, in part because we realize that birds are more closely related to them than was once thought.

Then there are those dramatic dinosaur roars. We all know that a big dinosaur sounded a bit like a bull elephant, don't we? Except we have no evidence at all to make this assumption. Dinosaur roars came out of science fiction, not paleontological fact. It has even been suggested that the most famous predator of them all, T. rex, was a scavenger. Over time, the accepted views of the life and times of dinosaurs, based on the gradual accumulation of better finds, has changed significantly, but remains speculative. No more so than with those scary Velociraptors from the movie.

Until *Jurassic Park,* most of the portrayal of dinosaurs in movies had concentrated solely on the big guys. Not surprising, given the impressive nature of the large-scale animals—that's what gets children excited in the first place. But the vast majority of dinosaurs were no bigger than chickens, and, as *Jurassic Park* demonstrates, particularly if they operated in packs as seems likely (this is still speculative), mid-range predators like Velociraptors could in some ways have been more of a threat to human-sized prey than the clumsy giants. Anyone who has watched the scene where the children hide in the kitchen as the Velociraptors stalk it can appreciate just how terrifying these things looked. So it's a bit of a shock to discover that Velociraptors almost certainly sported feathers.

Bone is by far the most likely material to survive the fossilization process. What typically happens is that the dead animal or plant lies in a pool of mineral-rich water, or in ground water. Over time, the less solid parts of the structure decay, and minerals are deposited in those gaps, filling them up. This can result in remarkable casts of delicate items like plants. Harder parts like bone and shells tend to go through a much slower replacement process where the original material is gradually eaten away and replaced by a different mineral, so that what was once bone is now rock. We tend to think of fossils always having this rock-like nature, but the term is used for anything that has changed form over time. Amber is a resin, but it is considered a fossil because it has undergone chemical changes from the hardened tree sap that it once was. Similarly fossil wood does not have to have been turned to stone or "petrified," though it often is.

Bone may be the most common part of an animal to survive as a fossil, but other materials do sometimes make it, and as paleontologists moved away from the obsessive hunt for big bones to a more scientific approach, we have discovered more and more ancillary remains, notably feathers. As we've already seen, dinosaurs were the ancestors of birds and a fair number of them do seem to have been feathered, including the Velociraptor. It is still likely that they were impressive hunters that you wouldn't want to meet on a dark night—but they were predators in comedic fancy dress.

Given the meager remains that are likely to remain intact from so long ago, it might seem that Michael Crichton, the author of the original *Jurassic Park* novel, had an insuperable task on his hands in justifying his plot. How could he come up with an even marginally plausible scientific mechanism to explain the ability to produce clones of something that hadn't been alive for so long? As usual, this was science fiction, not fact, so he could make some implausible, or even impossible, leaps, but the process had to seem feasible. What he came down on was really very clever—if, as we shall see, fatally flawed.

First, Crichton gives a false trail, commenting: "Of course, no dinosaur DNA was known to exist anywhere in the world. But by grinding up large quantities of dinosaur bones it might be possible to extract fragments of DNA. Formerly it was thought that fossilization eliminated all DNA. Now that was recognized as untrue." This seems an unlikely statement, given the nature of fossilization, but there is a real loophole, in that fossilization is not always complete. For example, in the 1990s a T. rex skeleton was discovered where some of the specimen was in a "relatively unaltered state," which enabled researchers to grid up bone and discover chemicals related to hemoglobin. But there is no way for DNA to survive in this fashion.

Then, Crichton gives us his "real" technique. His fictional expert Henry Wu explains: "Tree sap often flows over insects and traps them. The insects are then perfectly preserved within the fossil. One finds all kinds of insects in amber—including biting insects that have sucked blood from larger animals." It sounds plausible. Insects do regularly get trapped in tree sap, which over time becomes amber—and pieces of amber are found with insects embedded inside, sometimes from the distant past.

But the devil is in the detail, and in this case the problem is with that expression "perfectly preserved." An insect may well be preserved so that it looks much as it did when it died many years before, but that does not mean that the chemicals inside it fail to deteriorate. DNA is a very complex and relatively fragile substance. Once the cells containing the DNA die, enzymes can attack the bonds that make up the double-spiral structure of the long DNA chains, and water also degrades these bonds. Although there were some reports of extraction of DNA from ambers dating back around 130 million years made in the early 1990s, inspiring Crichton's fiction, they have proved impossible to duplicate and are considered to be a result of DNA contamination of the samples.

In 2013, researchers at Manchester attempted to reproduce the *Jurassic Park* technique, using insects embedded in copal, an intermediate stage between tree sap and the fully fossilized amber. This experiment was undertaken under full forensic conditions, unlike the 1990 studies. What's more, the new experiment was able to make use of more modern DNA amplification techniques, which are less likely to emphasize modern molecules than the PCR process used in the original studies. The experiment detected no DNA in the 60-year to 10,000-year-old samples, suggesting that insects trapped in amber lose DNA faster than those that have been air-dried.

Interestingly, it is possible to give DNA a "half life" like radioactive material, the period of time in which you would expect around half of the DNA in a sample to be destroyed, and this has been demonstrated to be 521 years. This figure was established by palaeogeneticists who examined a range of bones, aged between 600 and 8,000 years. While the actual lifetime of DNA will be influenced by a range of conditions, especially the temperature at which the sample is kept, this gives an absolute maximum beyond which no trace of DNA would remain.

The oldest DNA sequence to date comes from around half a

million years ago, while the half life suggests that DNA would be unreadable after around 1.5 million years and totally destroyed by a maximum of 6.8 million years. This leaves a huge gap to reach the 65 million plus era of the dinosaurs, and consigns producing a real *Jurassic Park* firmly into the bounds of fiction. Had decent DNA samples been conjured up somehow, the basic concept by which new dinosaurs would be produced, with a rich history in science fiction, is cloning. (The word "clone" comes from the Greek word for a twig, suggesting a branching off, producing multiple versions from a single original.)

Fictional clones often suffer from two distinct flaws. They are represented as exact copies of the original, and they can be produced immediately, or at least after a matter of hours or days. In the real world, a clone is indeed a kind of copy, in that it is based on the same DNA as its "parent"—but organisms are much more than a simple construction kit with DNA as the blueprint. In practice, the way that DNA is accessed during growth can produce very different outcomes, depending on the environment and how different genes within the DNA are turned on or off. We've all come across natural clones—identical twins—and they can be distinctly different.

The same has proved to be the case with man-made clones, since we have had the technology to make this possible, if still difficult. A good example was the first cloned cat, named CC. In practice it was anything but a carbon copy of its parent as it was black and white, while the parent was tabby. And like every real clone, CC was born and had to grow to full size at the usual rate. There was no overnight production of a "copy cat" in some kind of magical cloning chamber.

Once we understood the nature of DNA and the workings of cells it might seem that cloning would be a trivial exercise, but it has proved anything but. Cloning is reproduction not by the combination of half of the genetic material from each of two individuals, as in normal sexual reproduction, but by duplicating a single set of genes. When cloning an animal like Dolly, the sheep that was in 1996 the first mammal to be cloned, a piece of genetic material is taken from the original host, in the case of Dolly's parent, the source was from the mammary, which is why the lamb was named after the singer Dolly Parton. As it happens, Dolly's "mother/sister" was long dead—the cell used was from a culture, kept alive in the laboratory, not taken direct from a living animal. The contents of one of the donor cells are used to replace the insides of an unfertilized egg cell, which had its normal contents sucked out.

With a touch reminiscent of that great example of protoscience fiction, *Frankenstein,* a tiny burst of electricity was used both to help the nucleus fuse into the egg that would become Dolly, and to initiate the process of cell division. The egg, implanted in a host mother, began to grow in the normal fashion, and after the appropriate period of time, Dolly was born. She appeared to be a perfectly normal, healthy lamb. This description represents the simplistic version of history. If cloning were as trivial as that, we would have clones popping up all over the place, and the few individuals who say they have produced human clones would be proudly displaying them instead of making grandiose claims but never producing any evidence. In practice, getting that far has proved hugely difficult.

Even producing Dolly took many years. Although it had been possible to use this technique with frogs for some time, it just wouldn't work with mammals. The breakthrough was to start with a cell in a different state to those that had originally been tried. Most of the time our cells aren't rapidly duplicating themselves as they do in a fetus when it is growing. All the original experiments used cells that were already in the right state to split. But the cells used by the team at the Roslin Institute in Scotland who cloned Dolly were quiescent, cells that had initially been splitting, but then had their nutrients removed, so that the growth process stopped. Somewhat to their surprise, these proved effective.

The snags weren't out of the way yet, just because the cell had started to divide. Although the quiescent nuclei did seem to work when transplanted into an egg, most were false starts. Out of 276 initial tries, only 29 showed any sign of activation, and of those 29 implanted in surrogates, only one—Dolly—lived. But surely, now we've had Dolly, it's easy to get better and better at the cloning business? Isn't it only a matter of time before we can clone anything, dinosaurs included, if we have the right genetic material?

Not necessarily. It is certainly true that animal cloning has become more routine and reliable, but there are still issues. Clones' genes can get damaged in the rough-and-ready process of bringing a clone into existence. Cloning is a bit like trying to repair a delicate watch with a hammer and chisel—you can get lucky and fix things, but it's much easier to do damage. Later studies of animal cloning have shown that the process tends to modify the DNA, damaging important genes and resulting in the inability of many embryos to survive. Those that do live often suffer from serious problems. All the evidence is that these potential problems get worse with monkeys, worse still with apes, and it is quite possible that it may never be practical to safely produce a cloned human being without producing many damaged children. And, of course we know a lot more about humans and domestic animals than we do about dinosaurs.

Some have argued that all cloning will always be an uncertain process that throws up many failures. Certainly, the success rate so far is dismally poor with mammals—typically between

0.5 and 1.5 percent achieve a successful birth of an apparently undamaged baby animal—and although cloners are getting better at the mechanics of producing a cloned egg, they don't seem to be making much impact on this failure rate. Bearing in mind a good number of these failures take place after the embryo has started to form, or even after the animal is born, this is a totally inacceptable risk if the same technology were applied to humans. No one with any moral standing would consider it acceptable to produce a clone if the known price was the production of tens or hundreds of deformed fetuses and babies as by-products.

Some publicity seekers even claimed to have produced human clones, though strangely they have never been able to produce the very simple proof of the truth of their claims. This brings to mind the borderline science fiction of the 1970s movie from an Ira Levin novel, *The Boys from Brazil*. The borderline nature was that the science was just a kick-start for an action movie, but there are certainly some powerful science fiction themes in the notion of whether or not it would be ethical to make a human clone, especially when, as in the film, the subject is Adolf Hitler, raised in the hope of generating a replacement of the Nazi leader and bringing back the Third Reich.

If (unlike Hitler) the inability to clone dinosaurs is a disappointment, this doesn't mean that all extinct animals are out of reach. Mammoths make a particularly interesting potential target. They lived as recently as 5,000 years ago, recent enough both to have a decent amount of DNA surviving. What's more, they are regularly discovered preserved in peat bogs and frozen tundra, which gives them a higher potential for retaining DNA-carrying material that can be successfully recovered. Even better, we have a reasonably good surrogate in elephants—as another minor problem that *Jurassic Park* brushed over is that all successful cloning has involved the fetus being carried by a surrogate parent, and finding a match for, say, a T. rex to carry its egg to the stage of being laid could be challenging.

At the time of writing there are at least two projects aiming to clone a mammoth, Akira Iritani at Kyoto University, who has been working in the area since 2011, with a target of success by 2016–2017, and Hwang Woo-Suk, a South Korean ventinarian who recently set out on a crash course in mammoth recovery. The fact that Hwang is involved is not in itself very encouraging, as he infamously fell from grace as a noted stem cell researcher when he was dismissed from a post at Seoul National University in 2006 for having forged data from imaginary research into stem cells. Yet he certainly had experience of cloning, (assuming the experiment was not faked), when he cloned a dog in 2014.

Other scientists are both doubtful of the short-term chances of success, bearing in mind how many failures it took to get to the much less ambitious Dolly, and dubious that there is any scientific merit in bringing a mammoth back to life. After all, *Jurassic Park* was, in the end, a theme park, not a true scientific venture. While there would indubitably be some information gleaned from having access to a living mammoth, those who disagree with the process argue that any benefits are tiny and not worth the effort or the potential of causing the cloned animals suffering. As yet the mammoth DNA samples that have been found are not close enough to being intact to consider cloning, but this doesn't meant that the idea has not got potential.

Cloning isn't, of course, the only aspect of the biological sciences that has taken the fancy of science fiction. Thrillers involving escaped killer diseases, which tend to be genetically modified to be more deadly, are relatively common—and there can be no doubt that there are real-world concerns about pandemics being caused by escaped diseases and terrorist plots.

Here reality and fiction are only separated by the likeliness of a successful attempt and the technical details of the disease capabilities. Genetic modification itself has been a staple concern for science fiction authors since the 1970s. We tend to make a false distinction, but in fact human beings have been engaged in genetic modification for thousands of years.

If you doubt that, take a look at the range of dogs from Chihuahua to Great Dane that have all been bred from the same basic wolf-like stock, or at plants like maize and cauliflower that we have transformed so much that they are incapable of breeding without human intervention. Selective breeding is simply and certainly just as much genetic modification as anything we carry out today, though it is much more random in its outcome. The potential is for genetic engineering to enable much more targeted and specific changes.

Perhaps the closest we have come to science fiction in real-world genetic engineering, raising visions of H. G. Wells's dark tale *The Island of Dr. Moreau,* is the creation of human/animal chimeras where human cells and animal embryos are combined. Yet there is an important distinction that takes the impact out of the SF story line—where the fictional result is a monstrous crossover creature, in reality the result is not viable as a living organism, but results in the production of stem cells, which can then be used for medical purposes.

Wells envisaged bizarre hybrids like the leopard-man, the satyr-man, and the dog-man on Dr. Moreau's island. But he realized the impracticality of giving a human being wings. This hasn't stopped science fiction writers from exploring one of humanity's great dreams. What if the right technology could enable us to leave behind the airplane and soar solo into the sky?

7.

SUIT UP

At my peak of reading comics, I wasn't much of a Marvel fan. In the UK back then, comic books were seen as primarily for children, and to my ten-year-old self, the teen angst of the likes of Spider-Man was just no fun. I preferred DC Comics for their more straightforward, black-and-white view of the world. Superman's straightforward philosophy seemed to make more sense that the brawling, emotional aspects of Marvel's crew, while Batman was my favorite, back before the more recent reboot to his original dark persona. In part this was because he was just a human being—no special powers required which made playing Batman so much more feasible. And then there was the desirability of his utility belt, a surefire winner for a child.

Because of this, I didn't come across Iron Man until the recent series of movies. Of course, like the robots in chapter 5, Tony Stark, when suited up, demonstrates strength beyond human limits. Yet in reality Iron Man's most impressive feat is bringing to life a great human dream: the ability to fly. The science fiction in comics is, frankly, rarely as imaginative as the material that is found in a full-scale novel. It's not surprising—there just isn't the space to develop a speculative science theme to such a degree. But Iron Man plugs straight into a human dream, perhaps made even more human by the frequent technical problems that beset his suit. (We need to ignore the rapid assembly mechanisms used in the movies, which is pure fantasy—but flying is something of a quite different level of practicality.)

As with a number of the other key science fiction themes, the dream of flying is one that goes back all the way to Ancient Greece, to the story of Daedalus and Icarus. This pair made an escape attempt from Crete where they were imprisoned after Daedalus helped Ariadne guide Theseus through the famous labyrinth with its monstrous Minotaur. According to legend, the wings that the father and son team constructed were made of "wax and feathers." This seems extremely impractical even by the Ancient Greeks' decidedly fuzzy ideas of physics—but the implication seems to be that the feathers were attached directly to the flyers' arms using the wax. Icarus, of course, despite his father's warning, flew too close to the Sun, the wax melted, and he fell to his death in the sea.

Realistically, this is much more of a morality tale than it is any kind of practical flying mechanism. What the Greeks didn't realize (hence their propensity for dreaming up flying horses) was that birds did not make a good model for animal flight. When they scaled bird wings up to make the human or equine equivalent, they were unaware that a bird is extremely light for its volume, using cunning weight-reduction measures worthy of the designer of a supercar, like having hollow bones to reduce its mass. Combine this with a musculature that gives a very different force on the wing to anything human arms could provide and you have a proposition that simply can't work at a human size. Not that it stopped people trying.

The problem with emulating the wings of a bird is getting enough lift out of flapping. But some birds present a very different model. Watch a hawk soaring with wide open wings, and it hardly seems to move those wings at all. It manages to fly without making use of the effort that we can't summon up, gliding on updrafts, letting the air do the work. We might still have weight problem, but in glider form we can build much bigger wings with greater lift, supported by a more rigid structure. Surely then we should be able to fly? It might not provide the powered precision of Iron Man, but becoming a human hawk has a certain independent grandeur.

This seems to be the idea that has inspired a whole string of people who have been hurling themselves off buildings with makeshift wings ever since the Middle Ages. Admittedly some did try arm flapping with all the disasters that inevitably brings, but others went for the simple, rigid winged glide. There was plenty of experience here from the simpler flight device of kites, which had been around since the fifth century BC. While there are various apocryphal stories, often written long after the event in question, one of the more reliable accounts comes from the eleventh century, featuring a Benedictine monk called Eilmer, a resident of Malmesbury Abbey in England.

If the account of Eilmer's flight is true, he seems to have had a fixed wing on his arms and a tail on his feet to provide stability. He is said to have leaped from the top of a tower—probably that of the Abbey—and to have flown around 200 meters (660 feet), which would have meant he was in the air for about 15 seconds. The account from the early twelfth century suggests that he panicked and lost control, resulting in a crash in which he broke both of his legs. But he had achieved something close to the flight of a soaring bird.

While there is no doubt that many other attempts were tried in the intervening period, the breakthrough for winged flight as we now know it came at the beginning of the nineteenth century, when the English engineer George Cayley designed a glider to carry a human passenger, including all the basics of fixed-wing flying that come into play today. His designs were eventually put into practice, with first a boy as pilot and then a full-sized adult. Now, of course, wing suits provide the individual gliding experience in the most personal fashion, bringing us closer to the desired independence of flight that most of us can only achieve in dreams. But entertaining though this line of development is, it doesn't capture the thrill of Iron Man's ability to hover and maneuver effortlessly under power. For that, we need to follow the story of the development of the rocket belt.

Apart from the spaceship and the ray gun, there is probably nothing so archetypally sci-fi (the flashy movie version, as opposed to the more literary science fiction) than the rocket belt or jetpack. This is the Iron Man experience brought to life, echoing Superman's most striking superpower, once he made the transition from leaping over high buildings to soaring through the sky. It's the embodiment of that final scene from *The Matrix* that brings a big grin to the audience, despite its lack of originality, when Neo takes control and soars into the air untrammeled by the conventional paraphernalia of flight.

Admittedly, a rocket belt is anything but technology-free, but the simplicity of flying without wings or a vehicle, soaring straight up into the sky from a standing position, still gives that sense of freedom. Rocket belts were a central feature of the various incarnations of Buck Rogers, they helped the Jetsons commute, and, as we shall see, one appeared for real in the James Bond movie *Thunderball*. Iron Man's suit may be 90 percent fantasy, but rocket belts deliver on ability to fly. The thrill of the rocket belt was enough to encourage the great science fiction writer Isaac Asimov into the perilous field of prediction, suggesting in 1965 that rocket belts would be commonplace commuter wear by 1990. And of course, the military were all too interested in a device that could quickly and accurately position troops on the battlefield, or lift them instantly out of harm's way.

Although there may have been earlier attempts, the first widely documented work on a rocket belt comes from the early 1950s. An engineer called Thomas Moore built a prototype "jet vest" and tested it in tethered flight at Redstone Arsenal army base in Alabama (allegedly with support from Wernher von Braun, the man behind Nazi Germany's V-2 rockets and much of the early NASA rocketry). His $25,000 budget ran out before any more progress could be made. The same fate befell the rocketry engineering company Thiokol's attempts at producing a "jump belt" using nitrogen canisters as propulsion, a project that also ran out of cash as the military temporarily lost interest.

There was more success at Bell Aircraft Corporation, the company behind the X-1 aircraft that was the first to break the sound barrier. At Bell, engineer Wendell Moore (as far as I'm aware, no relation to Thomas) had developed small rocket thrusters, which were located on the X-1's wingtips and tail to give extra control at high altitude. Moore speculated that a pair of these thrusters would enable a man to "fly like Buck Rogers," and put together a jury-rigged structure he called the Small Rocket Lift Device. Like Thiokol's prototype, this used pressurized nitrogen to feed the rocket nozzles. In his test rig, Moore managed to reach 15 feet, flying in a (relatively) stable fashion while tethered.

By 1960, military interest was returning, and with it funding, enabling Moore to totally redesign the newly renamed Rocketbelt. While compressed gas was sufficient to test the principle of a personal rocket motor backpack, something more was needed to give the sustained power needed for a useful flight. Moore made use of hydrogen peroxide, a simple compound related to water with the formular H_2O_2 that is probably most familiar from hair dye. Hydrogen peroxide's structure is relatively unstable, breaking down to water and oxygen with the emission of heat. When the peroxide in the rocket belt's tank was forced over a silver catalyst it converted to high pressure steam, which was then emitted from the rocket nozzles to provide thrust.

The first true free flight came on April 20, 1961, when Harold Graham, a young engineer and rocket belt fan, took to the air from open ground near Bell's Niagara Falls plant in Buffalo. That first flight lasted for 13 seconds (pleasingly similar to Eilmer of Malmesbury) covering 112 feet at around 18 inches off the ground. Progress was quick, leading to flights as high as 15 feet, an outburst of public interest and even a demonstration for President John F. Kennedy. Bell Aircraft rightly felt that their Rocketbelt made for great publicity whenever it was put on show.

Despite occasional accidents, notably when Graham fell 22 feet and decided that rocket belt flights were no longer for him, Bell seemed to be going from strength to strength. To demonstrate that the rocket belt did not need experienced engineers or pilots to fly it, the company brought in a nineteen-year-old novice, William Suitor, who happened to earn a few dollars by cutting Wendell Moore's lawn. The company trained him up just as an opportunity came up to get the rocket belt in front of a worldwide audience. After a successful demonstration of the belt at the 1964 New York World's Fair, the makers of the James Bond movies got in touch. They wanted something new and dramatic for the iconic pretitle chase sequence for their latest film. In *Thunderball,* Bond was going to escape his attackers using a rocket belt, and Suitor stood in for Sean Connery as its pilot, as did senior Bell test pilot, Gordon Yaeger. The appearance in the movie was a great success, even though many assumed that the rocket belt was a special effect.

Despite many test flights at Bell, the military once more lost interest because little progress was being made toward practical battlefield technology. Rocket belts would continue to be used for publicity stunts and as entertainment—notably in the opening ceremony of the 1984 Los Angeles Olympic Games, flown by the now veteran Suitor—but they weren't practical transport devices. It's arguable that the rocket belt really hasn't moved forward much since the 1960s. Apart from stability issues, which could be solved by modern technology in the same way that the two-wheeled Segway personalized transport uses electronics to keep its rider stable, the big problem that seems impossible to overcome is the flight duration. Fuel is heavy and there is a limit to how much the pilot can carry—if you can't walk around in it, it's not a rocket belt: it's a small aircraft. Iron Man gets around this with a power source that is little more than handwaving magic, but in the real world, a rocket belt needs a source of energy. And that limits flights to a minute or two maximum. The rocket belt is, sadly, never going to help you to soar over the traffic jams on your commute to work.

More progress has been made on flying cars, which like rocket belts have had a long history of attempts and promises that haven't quite been fulfilled, but have come close now to commercial availability. Yet it is hard to imagine a workable city full of them, unless they were fully computer controlled. And, more importantly for our point of view, a flying car is just a special kind of light aircraft—it lacks the "flying person" factor. We need to return more explicitly to Iron Man for another of Tony Stark's special capabilities—as well as flight, the suit gives him extra strength and speed. And here, the reality is much closer to the fiction than is the case with rocket belts.

Special clothing that gives us the ability to walk faster and farther has a long fantasy tradition in the form of 7-league boots. The concept of 7-league boots was simple, even if the mechanics of using them was never clearly explained. A normal stride is around a yard. With these magical boots in place, somehow every step you took would cover seven leagues. A league was a distance of around 3 miles, so that gives you 21 miles for each pace—not bad going. If you took a sedate stroll at 1 step a second, this would amount to traveling at around 75,000 miles per hour. If it weren't for the sea, you could get all the way around the globe in less than 20 minutes. Of course, you would be traveling much faster than the speed of sound (which is around 700 miles per hour), so you would leave a sonic boom in your wake. (This assumes that you really had a 21-mile step at a normal pace. I've never been clear how the wearer of 7-league boots progresses. Perhaps the idea was more like floating over that 21-mile gap, so you would proceed in huge hops that took minutes or hours to complete instead of even paces.)

Seven-league boots may be pure fantasy, but there is now a way to extend your stride using a range of spring-loaded "jumping stilts" that strap onto the wearer's legs. With names like PowerStriders, Pro-Jumps, and PowerSkips, these allow the user to jump up to 6 feet in the air, and to run with 9-foot strides—not exactly seven leagues, but far exceeding normal human capability. The strap-on sprung stilts take some getting used to, but expert wearers can perform gymnastic feats that are impossible for the unaided body. In effect, these devices emulate the kangaroo. This animal once baffled zoologists as it seemed to use more energy in jumping than it consumed it food—but this was because its sprung rear legs absorbed energy as it landed and reused it, in the same way that the spring-loaded stilts do.

A company called Applied Motion have gone one step further (almost literally) than the strap-on stilts, building sprung legs into a frame that supports the whole body. The original version of the SpringWalker, a product that the manufacturer refers to as a "body amplifier," is powered by human energy, using the legs and arms, but the aim is to have externally powered versions. This development has been promised for a number of years, but at the time of writing there is no evidence of progress. Like jumping stilts, the SpringWalker gives the wearer the ability to move at an unusually fast rate, which in the powered version would also be effortless—and capable of covering a far more varied terrain than a wheeled vehicle.

Devices like the SpringWalker take reality closer to the most dramatic fictional equivalent, the military exoskeleton—huge, vaguely human shaped, machines that mimic the movements made by their pilot. Probably originally inspired by H. G. Wells's truly terrifying Martian fighting machines in *The War of the Worlds,* these are typified by the suits in games like Heavy Gear and *MechWarrior,* or the fighting machines in *The Matrix*. In essence, the SpringWalker is a simple, low-tech mechanical exoskeleton. There is nothing new about the concept of a creature with an exoskeleton. It's just an animal with a hard structure on the outside, rather than the inside where creatures like us keep our skeletons. There are many more animals on this planet with exoskeletons, from ants to lobsters, than there are with a backbone like ours.

The military exoskeleton has gone through an all too typical cycle in the real world, from enthusiasm for a paper dream to major modification when practical developments were tested. DARPA, the U.S. Defense Advanced Research Projects Agency, founded in 1958 as a response to the Soviet launch of the Sputnik satellite, originally intended to go far beyond the SpringWalker, not only extending the human stride, but expanding the whole physical capability of a human body, and doing so with serious horsepower, not just the puny leverage of a spring. DARPA ploughed millions of dollars into the development of military battle suits in its Exoskeletons for Human Performance Augmentation program.

DARPA's vision was to give a human being a metal external skeleton, which could be powered to provide immense strength and speed. This isn't the same as a cyborg, a machine/human hybrid (see page 207), but a powered suit, a lightweight version of the large-scale fighting machines of fiction. Unlike humanoid robots, much of the capability of an exoskeleton can be achieved with relatively simple mechanics (and a touch of hydraulics and gyroscopes, plus computer power to provide control and maintain balance, thrown in). DARPA's research and development did go as far as working prototypes. The aim, as much as possible, was to mimic the motions and capabilities of the human body, particularly in carrying and moving, but with strength or speed that far exceeds human capabilities. The most advanced of the models that were developed used a hydraulic system, employing a fluid link to amplify any inputs.

A simple example of a hydraulic system is the hydraulic press, which like many other water powered devices, was first used in the nineteenth century. The system works by enclosing a body of fluid in a cylinder with two pistons, one with a small surface area, the other much larger. Pushing the small piston into the fluid results in pressure on the large piston, which moves out as the smaller piston moves in. Because of the difference in volumes of fluid displaced, the large piston moves less far, but feels a proportionally larger pressure. So a small force on the small piston is turned into a large force (working over a much smaller distance) on the large piston. It's a fluid-based equivalent of gearing.

Such a system can be used to turn even manual effort into greatly increased force, but in the prototype exoskeletons, the source of the power was more likely to be either electrical or an internal combustion engine. The electrical option is the most desirable and flexible—there is something endearingly makeshift about having to start a gasoline engine on an exoskeleton as if it were a lawn mower or outboard motor—but the limitations in battery technology mean that an electrical exoskeleton, just like an electric car, would be limited in its capabilities.

An exoskeleton that ran out of power halfway across a battlefield would not be popular. DARPA's aim was to have a device that would "assist our soldiers in carrying [their load of armor, weaponry, and supplies] by developing a fully integrated exoskeleton system that will increase the speed, strength and endurance of load-burdened soldiers in combat environments." The idea was to have a heavily armored exoskeleton that was designed to have mission-oriented packages attached to provide the weaponry and capabilities required in different environments. When the exoskeleton was working, the operator would be able to move naturally, unencumbered, and without additional fatigue, while the machinery carried the payload.

One of the earliest examples of a practical exoskeleton emerged from the University of Calfornia at Berkeley. In 2004 their Robotics and Human Engineering Laboratory demonstrated BLEEX, the Berkeley Lower Extremity Exoskeleton, a pair of powered metal leg braces connected to a backpack that enables the wearer to carry heavy loads over long distances on foot. The initial version made the 100-pound device, plus a 70-pound load feel as if the wearer was carrying no more than 5 pounds in weight. After a number of years of further development, these heavy-duty suits seem now to have been mothballed for a more subtle approach.

The first of two current developments is TALOS—the Tactical Assault Light Operator Suit—which is a design specification from the Special Operations Command (SOCOM), requesting proposals for prototypes of what SOCOM itself has referred to as an "Iron Man suit," with the aim of having a product available in four to five years. This is intended to provide some of the facilities of the old-fashioned heavy suits, notably being bulletproof, carrying weapons, and increasing strength, but in a more lightweight and flexible form. Meanwhile, DARPA is looking at soft technologies, developing a suit system it calls "Warrior Web" that will help support the fighter using special fabrics to reduce fatigue and injury, rather than give any superhuman ability.

Although military exoskeletons, with their sci-fi connotations, tend to grab the headlines, there are also real possibilities of such soft exoskeletons being used to help the disabled where limb or nervous system damage makes it impossible for an individual to use their own muscles. In 2006, Panasonic demonstrated a compressed air-driven "power jacket," which amplifies the movements of arm muscles to help patients recover from partial paralysis. Unlike even the latest military exoskeleton concepts, the system is very light at only four pounds, though it does require a connection to an external compressed air device, and is more a treatment than an augmentation for everyday living—yet it demonstrates how this kind of technology can also have medical applications.

Unlike the military, Professor Yoshiyuki Sankai of the University of Tsukuba in Japan along with Daiwa House Industry Company and Cyberdyne have taken a form of exoskeleton all the way to a released product. Although it arrived in 2013, well after the original target of 2008, their "Robot Suit HAL" provides a strap-on support to help those with limited physical ability. At the time of writing several hundred suits are in use, and a modified version went into use in 2014 to give super strength to building industry workers. One of the clever aspects of the design from the very first was to use detectors on the skin to pick up the signals that the nervous system sends to the muscles as the way of controlling the suit. Where the fictional exoskeleton usually responded to actual physical effort from the person inside, the medical application needs to be able to work even if the user is unable to make the appropriate movements.

One of the problems with using electrical motors, hydraulics, or compressed air to enhance the human musculature is that the mechanisms can be bulky, heavy, and take up a lot of room. The Panasonic jacket might be lightweight, but it had various tubes sticking out of it and required connection to relatively large pump system, while HAL requires a 22-pound battery. Researchers at the University of Texas at Dallas in Richardson have been taking a different approach, aiming to construct an artificial equivalent of the muscles in the human body.

Two early versions of an artificial muscle up to a hundred times stronger than the natural version have already been built. One is powered by alcohol. The "muscle" is a length of wire with a special catalytic coating that causes the fuel to burn flamelessly, heating the wire, which contracts, producing a pulling force. The second uses carbon nanotubes—extremely strong and thin tubes made from lattices of carbon atoms, which change shape when an electrical charge is passed through them. The carbon nanotube "muscle" works without a battery, as a catalyst is used to produce a charge from a reaction when oxygen reacts with the fuel. What is particularly clever about these artificial muscles is that they are both the actuator that makes the movement happen and the power source—just like a physical muscle—so they are much more compact than conventional mechanical devices.

In the medical applications it's clear that an exoskeleton is built around a human to support their bodies, but with the military applications, it is possible that the move away from constructing heavy battle gear is the realization that there is no great benefit to be had from having a human onboard. It's not that the future equivalent of fighting machines would necessarily become robots; they are likely to become the walking equivalent of an airborne drone. Why put a human being at risk, if the powered suit can go into action on its own, controlled by a remote operator many miles away? But however they are controlled, various forms of exoskeleton seem certain to continue to be developed for both military and civilian applications.

As well as the impressive flying powers, Iron Man has a beam weapon in his gauntlets. But death rays, ray guns, and phasers were staple weapons in science fiction long before Tony Stark took to the screen. It's time for a touch of zap.

8.

RAY GUN READY

When I was very young, playing outside (which we seemed to do far more than children do these days) would almost always involve at some point a battle between cowboys and Indians. No self-respecting kid would go out to play without carrying a toy gun, preferably one that fired "caps"—tiny circles of gunpowder in a paper strip that made an impressive cracking sound when the gun's hammer hit them. It didn't matter that the whole concept of the American West was a world away in time and space to a twentieth-century industrial town in the North of England, cowboys and Indians was part of our reality. But over time, the weapons of choice and the scenario changed. As science fiction began to become the preferred option in my entertainment world, the fake Colt .45 was discarded without a second thought for the toy ray gun, complete with flashes, strange noises, and sometimes even real sparks.

Captain Kirk, of course, had a lot to answer for. It's true that *Star Trek* was nowhere near the first to make use of beam weapons. Ray guns were a staple of some of my favorite sci-fi reading from the heydays of the 1950s, and were in use as far back as the 1920s. For that matter, movie and TV science fiction rapidly made everyone at home with the "blaster" or whatever the name of the month happened to be, but somehow the phaser used by the crew of the *Enterprise* took hold of the imagination in a way that the other futuristic weapons hadn't. (This was particularly true, oddly, when the design changed from its original classic ray gun shape to something closer to a pocket flashlight.) The phaser made the beam weapon as ubiquitous as Western TV shows had the six-shooter.

You could argue that the first death ray came not from science fiction but from reality—despite this being the strange form of reality of the Ancient Greeks that mingles with myth sufficiently that we can never be sure for certain whether the story has any historical basis. This earliest of beam weapons involved nothing more than the light of the Sun—but as anyone who has ever played with a burning glass knows, that can do plenty of damage with the right technology. The inventor of this hypothetical weapon was the Ancient Greek mathematician and engineer, Archimedes.

Born in 287 BC, Archimedes was fascinated by light and mirrors and wrote a book on optics that, like many of his works, has since been lost. Based in Syracuse, on the island of Sicily, Archimedes lived through a time of huge upheaval for Greece. The Romans, previously dismissed as barbarians with nothing to offer by the older Greek civilization, were attacking Greek cities across the established world, demonstrating that civilization wasn't always the best way to win a war. By 212 BC, Syracuse was in the Romans' sights. And just as the cream of American and European scientists were pressed into service to support the Manhattan Project during the Second World War it seems that Archimedes was called on to employ his engines of war in defense of the city.

There seems better evidence that Archimedes genuinely did work on mechanical devices, built to grapple and overturn attacking ships as they attempted to dock, but he was also said to have worked on a ray weapon, employing the concentrated light of the Sun to blast the incoming ships. Archimedes would certainly have known that curved mirrors could be used to focus sunlight to create a burning effect and he is believed to have drawn up plans to scale these up to form weapons that were to be deployed in banks by the sea, working together to focus the Sun's rays onto attacking ships. The hope was that the burning rays would start fires on warships made of wood, which often were made waterproof at the joints by coating them in highly flammable tar.

Realistically, it would have been difficult to have set the planking of ships alight this way. The targets would constantly be rising and falling with the motion of the sea, and wet wood is difficult to set alight at the best of times. But it is possible that that pitch-based sealant, or tinder-like items scattered about on the deck might have been encouraged into flame. Across a distance of over 2,000 years, no evidence remains that the mirrors were ever built. It's entirely possible that this idea seemed one step too far for hard-pushed craftsmen, whose time would be considered better spent on conventional weapons. The very concept of a device that could set things alight at a distance would have seemed alien in a simple world where physical contact was the only way to make things happen.

It is also possible that the idea was not fully developed. The mirrors could have been the project that Archimedes was working on when he died. According to legend, he is said to have been poring over diagrams in the sand when the Roman soldiers arrived in Syracuse. Without looking up, the story goes that Archimedes told a soldier to get out of his light and the invaders, not realizing who he was, are said to have slaughtered the insolent old man for his disrespect.

The regulars of the TV show *MythBusters* have attempted to reproduce Archimedes' mirror weapon several times and have failed to set a ship alight on each occasion, declaring the myth "busted." But a number of academics have cast doubt on the TV presenters' attempts, pointing out limitations in the way that the mirrors were aimed and in the construction of the reflecting surfaces. Other experiments away from the cameras have certainly produced flames at a considerable distance. It seems likely that the potential for success of this type of "ray gun" was borderline, and only ever would have been useful somewhere with pretty much guaranteed bright sunlight. Even so, everyone agrees that a concentrated light beam weapon would have caused confusion for the invaders and could have temporarily blinded the incoming crew and so could well have still been valuable in the defense of Syracuse.

To get closer to phasers, the concept of "ray guns" clearly doesn't go back as far as Ancient Greek times, but it brings together the twin strands of scientific and fictional development. The term "ray" was originally used to describe a narrow beam of light and goes back at least to 1400 in English ("A crystal clyffe ful relusaunt / Mony ryal ray con fro hit rere," which roughly translates as "A crystal cliff that shone full bright / Many a noble ray stood forth" in the fourteenth-century poem *Pearl*) and occurs as early as the twelfth century in French, derived from the Latin term for such a narrow beam, "radius."

By the nineteenth century, that versatile word was starting to be used to describe all kinds of insubstantial emissions. The astronomer William Herschel discovered infrared light while experimenting with the influence of various parts of the spectrum on a thermometer. Discovering there was more heat toward the red end, he continued to move the thermometer and found to his amazement that the heating effect of the light continued, getting stronger, well after anything was visible. He called this new kind of invisible light "calorific rays." Later there were alpha, beta, and gamma rays identified as coming from radioactive sources and cosmic rays from space. (In practice, only gamma rays would turn out to be true rays of light, the rest being energetic beams of particles.) And then there was the immensely popular discovery of the X-ray, a ray with the truly science fiction–like quality of seeking through solid objects.

Meanwhile, fictional weapons using focused light to cause havoc had reemerged from their long sleep since the time of Archimedes. In Washington Irving's 1809 book *The Conquest of the Earth by the Moon*, there was the speculation that technologically advanced aliens could be "armed with concentrated sunbeams," while the 1898 classic from H. G. Wells, *The War of the Worlds*, featured walking machines that could emit heat rays (presumably infrared) from an internal generator, using a special polished parabolic mirror. In the same year, a rapidly written imitation of the Wells book, *Edison's Conquest of Mars* by Garrett P. Serviss featured the impressive sounding "disintegrator ray." By the 1920s, ray guns and blasters had become commonplace fare in both novels and movies.

Perhaps the single individual who best straddles reality and science fiction when it comes to the death ray is Nikola Tesla. There has been more nonsense written about Tesla than pretty well anyone else in the history of science and technology. On the Internet you will find long screeds that exult Tesla as an unrecognized genius who invented practically everything imaginable, only to have nasty, money-grubbing rivals push him aside to claim the glory. Often poor old Thomas Edison is used as the bad guy to contrast with Tesla's sainthood.

The reality, as is often the case with actual people, as opposed to the fictional characters we try to turn them into, falls somewhere between the extremes of perfection and failure. The truth is that Tesla, born in what is now Croatia of Serbian parents, and resident in the United States from 1884, was a brilliant engineer but also often proved to be a terrible businessman, a poor scientist, and an inveterate fantasist. Because it's easy to blur the lines between science and engineering, it's easy to forget that the two are not synonymous. Some superb scientists have been absolutely hopeless at doing anything practical, while engineers often manage to do remarkable things without truly understanding the science that lies behind what they are doing.

Tesla did superb work on the basics of the alternating current (AC) electricity supply that has become the standard around the world. He was one of the people to first come up with the idea of using multiphase systems, so important to industry, and he devised the best AC motors of his day. Tesla also developed an innovative way to produce extremely high voltage, the Tesla coil, and some clever radio control and generator designs. Where things start getting messy, are over claims that Tesla was a world-leading scientist in radio communication, the discoverer of X-rays, and far more.

The evidence is very strong that Telsa did not understand electromagnetic radiation, and his ideas for world communication (and power transmission) systems that the conspiracy theorists believe were buried by more political players like Edison, depended on an unrealistic idea that you would be able to send electrical *vibrations* through the earth and that somehow radio waves would then make up a return circuit (an idea that is necessary for the electrical circuitry with which Tesla had such expertise, but that is not needed for radio transmission). Tesla fans paint him as a maligned hero whose ideas were constantly stolen by people like Guglielmo Marconi, who was then selling a product far inferior to Tesla's own. But what Tesla actually did was to promise to deliver systems over and over again that never worked and never would work, because he didn't understand the underlying physics.

Despite Tesla delivering clever and innovative engineering designs until late in life, these were usually in his mind stopgaps to bring in a little money before he could make one of his truly earth-shattering inventions work. He would continue to promise new wonders and yet deliver nothing. And it seems that Tesla's "teleforce weapon" or death ray fell into exactly that category. To be fair, Tesla seems to have objected to that term "death ray," which was given to his concept by the press. What he had in mind was not a light-based weapon, which he believed would lose energy too rapidly, but a device that was supposed to project charged particles at high speed.

With typical bravado, Tesla claimed that his particle gun would "bring down a fleet of 10,000 enemy airplanes at a distance of 200 miles . . . and cause armies to drop dead in their tracks." His design seems to have been like the ion thrusters now used on a regular basis in spacecraft, producing charged particles and sending them out in a collated stream by electrostatic repulsion. Of itself, this is perfectly possible, but the claims that were made for the weapon's capabilities far outstrip anything possible in reality, as perhaps Tesla would have realized if he had truly, as he claimed, demonstrated and used the device. This is why what is now a perfectly well-understood technology is not deployed on the battlefield, despite plenty of research into charged particle streams as weapons.

There certainly are firearms that make use of electromagnetism as an accelerative force, notably the rail gun, which uses electromagnets to accelerate a metal- or plasma-based component along a rail. A rail gun can accelerate a projectile to similar velocities to a traditional gun, but it bears no resemblance to a phaser. Closer perhaps, certainly in name, is the taser. That name is clearly based on "laser" (or even phaser) and was apparently an acronym for Thomas A. Swift's Electric Rifle, named after the hero of a series of juvenile science fiction books.

The best-known form of the taser that has been put on sale shoots a pair of barbed electrodes into the victim, then sends a high-voltage charge down a pair of wires that link the electrodes to the gun, disabling the victim. The manufacturers TASER International do already produce a wireless version, where a shotgun is used to propel a special baton round that contains a high-voltage battery, but this bears little resemblance to the ray guns of fiction: it is still a projectile weapon.

To all intents and purposes *Star Trek*'s phaser was a laser, though it suffered from the usual distortion of reality seen when science fiction TV and movies represent powerful light rays. If you watch an episode of *Star Trek*, the phaser beam takes a detectable time to get from source to target—in practice light is so fast that any short range is effectively crossed instantaneously. It seems that Gene Roddenberry made the subtle change of name from what was at the time still a relatively new invention as the plot required the phaser to have a "stun mode," something that there didn't seem an obvious way of doing with a conventional laser, so the name change was to avoid complaints from technology nerds.

Roddenberry seems to have been less concerned about the scientific inaccuracy of another characteristic of phasers and practically every other light-based weapon in fiction—that it can be seen from the side as a glowing beam. Psychologically this seems to have been important to distinguish the beam weapon from a projectile weapon like a traditional handgun. We don't expect to see the bullet on its way to the target at hundreds of miles an hour. But if a wound appears suddenly in a victim, it is too closely associated with the use of a bullet. The trouble is that light is not visible in the same way that a physical object is. When we see, for example, a pen, light in the environment bounces off the pen and reaches our eyes. But a beam of light, laser or otherwise, does not reflect light, so we can't see it sideways on.

We are, of course, familiar with laser-light shows where the beams are made visible, but this is only possible because the special effects people are pumping a considerable amount of smoke into the atmosphere. Some of the light in the beam is scattered by the particles in the smoke, reaching our eyes, so we see the beam. But that smoke isn't usually present when lasers are portrayed as blasting across space or down the corridor of a starship. The beams should not be visible from the side, any more than the beam of a laser pointer is when it is used to highlight an object on a screen. This reality has been accepted for laser targeting—we don't expect to see a beam when an ominous red dot appears on someone's chest in a movie—but it still isn't accepted for laser-based weapons.

Even so, there is no doubt that the laser is the closest that we have come to a real-world development of science fiction's weapon of choice, something that was made clear from the very early days of its development. The laser's immediate ancestor was the maser, an acronym for Microwave Amplification by Stimulated Emission of Radiation. This was the realization of a concept that Einstein had described in 1916. With the right material, electrons could be pushed up to a semi-stable, high-energy state by the passage of suitable photons of light. A second photon interacting with the electron would cause a pair of photons to be emitted, stimulating it to emit radiation. The result was a kind of light amplifier.

In the 1950s, this process was made real with ammonia gas, sending microwaves through the gas to build up to an amplified result. This was pretty much simultaneously announced as a theory by Russian scientists Alexander Prokhorov and Nikolai Basov and practically by the American Charles Townes, who built a maser before the Russians published their paper, but had yet to announce his work when his rivals went public. Masers were interesting, and had real applications in fields like telecommunications and the design of atomic clocks, but they didn't have large-scale practical applications. The idea of an equivalent using visible light was much more appealing and a whole host of scientists across the world would begin work on an optical maser, producing an unprecedented technological race.

Townes was once more involved, with a colleague at Bell Labs called Art Schawlow, but they had serious competition in their research from the younger Gordon Gould. Like the Bell team, Gould came up with the idea of using a pair of mirrors at each end of a cavity containing a suitable material, which would allow a beam to build up until it was intensely concentrated. Townes was coming at the research from the Bell viewpoint, looking for a better way of producing a communication signal, so was not particularly interested in taking this concept to the extreme, but Gould believed it was possible to create a beam that would produce the kind of conditions that are experienced on the surface of the Sun.

Gould was aware of the competition and wrote up some notes which he headed "Some rough calculations on the feasibility of a LASER: Light Amplification by Stimulated Emission of Radiation"—the first use of this term. (Townes, anxious to link the new idea to his existing device insisted on using the clumsier term, optical maser.) Unable to develop his idea further alone, Gould took it to a company called Technical Research Group (TRG), which specialized in defense contracts. When the company took the idea to the newly formed Advanced Research Projects Agency (ARPA) Gould proved a powerful advocate.

Rather than cautiously undersell his idea in a way that is more typical of a scientist, Gould dreamed up a string of potential military uses. The laser could, he said, send a communication beam to Mars, punch holes in metal, shine a beam on a target to direct incoming weaponry, or even destroy missiles with its remarkable power. If this sounds like Tesla's extravagant claims for his particle weapon, Gould had much better science to back up his claims. The result was a remarkable outcome. TRG was asking ARPA for $300,000 to develop the laser. Instead they were awarded nearly $1 million—a considerable sum at the end of the 1950s.

Inevitably ARPA wanted the project to be classified, and this is where the development of the laser takes on a tragicomic air. Gould had dabbled with left-wing ideas in his youth and as a result was refused security clearance. Two of the reasons for this, reflecting the very different times back then, were because Gould had lived with his wife before they were married, and because two of his referees had beards, making them potentially subversive. Gould was officially prevented from entering the building where his research was underway, and was no longer allowed to read his own notebooks, which were confiscated.

The delays that Gould was suffering due to his security clearance problems were matched by technical issues in the Townes camp, allowing a third key player to enter the race, the Hughes Aircraft Company engineer and physicist, Theodore Maiman. With considerable experience working with rubies in masers, Maiman decided to give them a go in a laser, despite the Bell

Labs team having declared that rubies would be useless, after some faulty research work by a third party that Schawlow took as correct without checking.

As often happens in such a development under pressure, it was a lucky break that gave Maiman the lead. Gould was bogged down by his security issues, Bell Labs had decided to follow the trickier gas laser route, leaving Maiman with just one major problem—how to get a bright enough light to stimulate his ruby, without overheating and damaging it as arc lamps would. Up to this point everyone had assumed that a laser would work continuously (as many of them indeed do), but Maiman's assistant, Charlie Asawa, also looked into the possibility of using short pulses of light, which gave the lamp a chance to cool and would prevent overheating.

As it happened, one of Asawa's friends was an amateur photographer who bought all the latest equipment. He had just got hold of a replacement for the painfully antiquated flashbulbs still used back then in low-lighting conditions. Flashbulbs produced an intense flash by overloading and burning out a magnesium or zirconium filament. The bulb had to be replaced after every exposure, making the approach clumsy and slow. Asawa's friend had got one of the first electronic flashguns, a device that would produce repeated flashes from the same tube, by sending a high-voltage pulse through low-pressure xenon gas.

In the early days of electronic flash design, the flash tubes were often spirals—Maiman made use of the shape to put a cylinder of ruby into the middle of a flash tube. One end of the ruby had a mirror, the other a mirror with a small hole to allow the pulse of light to escape. With the whole thing wrapped in aluminum to reflect back any light that escaped, it looked like something from a mad scientist's lab. But on May 16, 1960, this small, fist-sized device would produce the first laser light.

A few weeks later, Hughes Aircraft arranged a press conference to discuss the important potential applications that the laser would have in telecommunications. But the assembled press had different ideas. They knew what they were seeing and demanded to be told more about this prototype ray gun. Maiman, lacking Gould's flair for drama tried to deflect them from what he regarded as a silly idea, though he couldn't deny that the laser had the potential to be used as a weapon. To Maiman's horror, next day's headlines included "L.A. man discovers science fiction death ray."

No one could sensibly deny that Maiman had got there first with a working laser (Bell Labs tried, arguing for some time that using a ruby was their idea despite having dismissed the possibility out of hand), but it took much longer to establish who had priority in the theory. Gould, still without a security clearance that he never would receive, did eventually get patent recognition for his invention, but only in 1964 in the UK and in the 1980s in the United States. To rub salt into the wound, it was Townes, Basov, and Prokhorov who were recognized by receiving the Nobel Prize. Neither Gould nor Maiman got a look-in.

Four years after Maiman's first demonstration, the laser took over as the fictional ray gun of choice, when a (fake) bright red beam, originating from a device that was clearly based on the publicity stills from Maiman's work, was aimed at James Bond in *Goldfinger* with the intention of slicing the secret agent in two. Now, of course, we do use lasers for precision cutting and medical applications, but on the military side they have not made it into the armory as a sidearm. We are all familiar with laser targeting from the red dots of assassins in movies, but the closest to a handheld ray gun has been the development of laser weapons designed to blind an opponent. In practice these haven't been developed to any extent since a UN protocol was introduced that prevents weapons that cause permanent blindness being used. Although the prototype weapons were only supposed to cause temporary blindness, it is very difficult to be sure that the disabling effect won't be permanent.

Larger, ship-based lasers have also been developed, at least to the test level. Ships have always suffered from stability problems as a result of the recoil of guns (which is why rockets, which don't have recoil, were first developed militarily for use on ships), so they would provide an ideal base for a laser weapon. These devices are not envisaged as being direct, sci-fi–style death rays, destroying incoming planes and missiles by disintegrating them, but instead a way of causing sudden heating on the skin of a target, causing stress failure or damaging shock waves. Lasers have also, of course, come into our lives in a big way thanks to the tiny, solid-state lasers that are used in laser printers, CDs, and DVDs. We also come across the descendants of Townes's gas lasers in supermarket scanners. But as yet the laser has not become anything close to a classic handheld ray gun.

In 1977 (still only seventeen years after the first laser was built), science fiction brought us a different take on the death ray, crossing it with the sword, the fantasy world's weapon of choice, in the light saber. These devices wielded by the Jedi in the Star Wars series are supposed to set their bearers apart from the usual, blaster-firing crowd, making them both slightly antiquated and mystical at the same time. But they also bring in a particularly unusual—in fact, some would say ill-thought-out—twist on the physics of light.

There is one prime difference between a beam of light and a piece of metal as the material from which to construct a sword blade: light does not stop of its own accord. For a light saber to come out of its handle for a few feet then suddenly stop, we would expect there to be *something* there at the end—something from which the light could reflect back. But there isn't anything. No doubt there are after-the-event justifications of light-saber technology that describe it as some form of plasma beam, limited by a force field, but there is little doubt that the initial intent was a laser that just stopped in midair somehow. Everyone who knew anything about physics knew they were impossible. So it was quite a shock when, in 2013, newspaper headlines proclaimed that light sabers had now been made possible.

"Star Wars Light Sabers Finally Invented," "Scientists Finally Invent Real, Working Lightsabers," and "MIT, Harvard Scientists Accidentally Create Real-Life Light Saber" were among the dramatic headlines. (I love that use of "Finally" as if it is about time that those lazy scientists managed to get around to something so trivial.) This media frenzy was set off by one of the scientists behind the discovery, Professor Mikhail Lukin of Harvard, who made the, perhaps unfortunate, remark: "It's not an inapt analogy to compare this to light sabers." He was, no doubt, encouraged by the Harvard PR department. It all sounded very impressive. And it was an interesting bit of science. But the link to light sabers was tenuous in the extreme.

By using a special, low-temperature material called a Bose-Einstein condensate, the team had been able to produce what they called "light molecules"—pairs of photons of light that were temporarily linked to each other. This is an important breakthrough because photons generally don't interact with each other—so any mechanism by which they can be made to work together could be extremely useful when building devices that use light instead of electrons, so-called "photonics," that could eventually revolutionize computers and other devices based on electronics.

By forming a "Rydberg blockade," specially excited atoms that prevent photons from progressing, the scientists managed to get the first photon to hold the second photon up briefly, so the two became linked together, pushing and pulling through the medium as they interacted one after the other with excited atoms in their path. It genuinely is interesting stuff. But it has no relevance whatsoever to the practicalities of building a light saber. Unless we have a much more dramatic breakthrough, light sabers will remain stored away in the box labeled "Fantasy." And even the humble handheld ray gun is unlikely to make an appearance any time soon.

This leaves us without useful handheld beam weapons, but ray guns have traditionally also been the weapon of choice of another science fiction favorite. It's time to encounter the alien.

the first season, what he faces is essentially a human with a mask and claw gloves. (This encounter is often cited as one of the worst fight scenes ever in a science fiction production, and is hard to watch without sniggering.)

Like the transporter on that show, the human-form alien structure was adopted primarily for budget reasons, but by the arrival of *Star Trek: The Next Generation,* the writers felt it necessary to explain the very limited range of alien formats by retrofitting a backstory where all these human-like beings had a common ancestor.

We certainly can make some deductions about aliens from the diversity of life on Earth. Here we find relatively few basic "prototypes"—the variants of four limbs and a tail are found in pretty well all animals with backbones. So we might expect a different, but similarly alike-to-each-other set of prototypes on an alien world. We also have examples on the Earth where evolution has independently come to roughly the same solution to a problem from different starting points—so, for instance, an octopus and a human have similar-looking eyes, despite having reached them by wholly different evolutionary paths. But then there are other types of eyes—the compound eyes of insects or the stalked eyes of lobsters work very differently—so we might expect different organisms on a planet where light is available to develop a range of types of eyes, but we would expect eyes to be common, particularly in a species that has managed to develop to human-like intelligence or greater.

A very basic aspect of science that some science fiction designers could have paid more attention to is the problem of scaling. This is a simple physical reason that explains why we will never see giant spiders the size of those in the Harry Potter books or *The Lord of the Rings*—or for that matter, huge versions of human beings or aliens with similar body proportions. If you imagine blowing up the body proportions of any animal, the cross section and strength of the legs goes up with the square of the increase, because we are dealing with a two-dimensional feature. The mass of the body, by comparison, goes up with the cube of the increase, because here there are three dimensions. So before long the sheer weight of a giant spider, which is growing far faster than the cross-section of its limbs, would be enough to snap its legs like twigs.

Even Godzilla has serious scaling problems. In the original 1954 movie, this dinosaur-like monster was 50 meters (160 feet) tall. But with time he has grown bigger, reaching a massive 106 meters (350 feet) in the 2014 movie. Compare this with a Tyrannosaurus rex, which would have maxed out at around 12 meters (40 feet) long, including tail, and stood at less than half that in height. The biggest dinosaur discovered to date, found in Argentina in 2014, would have stood around 20 meters (65 feet) high—but this was a dinosaur like an Apatosaurus (the dinosaur formerly known as Brontosaurus) in shape. Despite its huge legs, Godzilla's bones would not have stood up to the strain of its mass.

The same goes for any conventional creature blown up in size, and quite often, aliens have been depicted as little more than enormous versions of lizards or insects. Technically, Godzilla isn't an alien, but an ancient sea monster, often described as activated by nuclear radiation. The same goes for the giant ants in the otherwise impressive sci-fi B movie *Them!*. They too are supposedly mutants rather than aliens, and similarly would have collapsed under their own weight. It might seem strange to complain when the dinosaurs Godzilla was based on might not have been *as* big, but were still vast. But creatures of that scale have legs that are much thicker in proportion than they are for a smaller equivalent (compare an elephant with a mouse for a living example).

So, extremely large aliens would need much chunkier limbs than they are usually portrayed as having. One possible way around this would be if the aliens didn't have the same composition as us. It is possible to imagine that they could have skeletal structures based on much stronger equivalents of bone, or could be silicon-based life (see page 123) with a greater ability to withstand the pull of gravity, perhaps needed for a high mass planet.

Gravity itself is a variable that needs consideration when we come to aliens. Could an alien life-form develop on planets with a much higher gravitational pull than Earth—probably requiring a totally different approach to movement, as legs would inevitably suffer—or in gravity fields so weak that they could float in the atmosphere like the balloon creatures in Ray Bradbury's *Martian Chronicles*? Might they be unbelievably delicate structures that would collapse in an instant under Earth levels of gravity?

The ultimate extreme for ultralight aliens would be for there to be organisms that had no mass at all. Like human-shaped aliens, this has been another favorite alien form on *Star Trek* over the years, as the "pure energy" being doesn't involve any expensive special effects apart from, perhaps, some twinkly points of light and a disembodied voice. The idea that aliens could exist as pure energy without a physical body was not uncommon in early written science fiction, but seems to have become less frequently used in recent years. This may, in part, be because of the similarity of the idea to the religious concept of a disembodied soul, which perhaps is less acceptable to a more skeptical modern audience.

The bigger issue is probably the "science bit," as they say in the cosmetics commercials. It is difficult to envisage how a living creature made solely of energy could exist as a physical entity because it would need some kind of retaining mechanism, perhaps a little like the force fields of chapter 4. This would have to be incredibly complex, because somehow it would have to store the memories and provide the brain and senses for this formless energy. It is hard to imagine a mechanism by which such a thing could evolve. Even if the aliens, as often was the case in early fiction, were once corporeal and had used technology to become pure energy creatures, there is no obvious mechanism to allow for all the requirements of a brain and ability to communicate and interact with physical objects to be transferred to an organism with no matter involved. It's one of those handwaving ideas that make no sense when you get down to practical detail.

Admittedly, we can envisage computers where the wiring is replaced with pure energy—beams of light—which have the advantage of being able to pass through each other unaffected, packing as many connections as you like into a limited space. But one of the essential aspects of light is that it needs to interact with matter for changes to happen—otherwise, it carries on regardless and unaffected. Without physical matter to constrain it and provide storage and processing space, the energy being would radiate outward and dissipate in an optical sigh—a romantic-sounding concept, but hardly a useful basis for an imagined life-form. A being that was a mix of energy and matter components would be feasible, but it's hard to see how a pure energy creature could exist.

Another popular exotic alien form in the early days of science fiction that took some notice of science involved life that was based on silicon, rather than carbon. Silicon occupies the position below carbon in the periodic table and so has similar bonding capabilities for its compounds, based on the same, flexible, four-bond structure. So it seemed reasonable to earlier

SF writers that you could have "stone life" in which silicon substituted for the universal carbon that is found in terrestrial organisms. Unfortunately, silicon has a significantly bigger nucleus than carbon, which prevents it from making the essential double bonds as easily and prevents it from forming the same bewildering variety of long chain molecules that carbon does.

It is this unique versatility and the ability to construct massive DNA and RNA molecules, amino acids, and proteins that seem essential for life. Human chromosome 1, as we have seen, is a single molecule of DNA that is by no means the largest chromosome in earthly species, yet contains around 10 billion atoms. While it is possible to conceive of life based on more and simpler molecules, there is still a need to produce sufficient complexity to deal with all the complications of biological life. Sometimes silicon does have equivalents to important carbon compounds, such as silane, the silicon equivalent of methane, but they tend to be very reactive with water, making them far less useful in forming a silicon equivalent of organic chemistry. Stone organisms seem unlikely ever to be a reality.

It's notable that we do find a lot of the basic compounds featuring in Earth-based living creatures, like amino acids, cropping up out in space and in comets and other debris—but we don't see any equivalent compounds with a silicon base. A handful of natural silicon molecules have been discovered, but far fewer and far less complex than the naturally occurring carbon molecules. And on Earth itself, though silicon is far, far more common than carbon, with around a thousand times as much silicon in the Earth's crust as there is of the lighter atom, it is carbon that turns up in living creatures over and over again. (In practice it's the carbon in the atmosphere that mostly ends up in living things, but silicon is still more widely available, and is likely to have been used had it been feasible.)

Whether or not alien life-forms are carbon based, science fiction, and particularly sci-fi, has largely portrayed the alien as our natural enemy. Of itself, there is no need for an alien species intelligent enough to cross the vast distances of interplanetary space to be belligerent, but from the early days, like Wells's ruthless Martians, they have been portrayed as deadly monsters. Again, evolution is to blame, as the evolutionary concept of natural selection seems to have been the main driver here. This was "nature, red in tooth and claw" taken up a gear, where technologically advanced aliens would not come seeking friendship, or a cultural exchange, but to dominate and take over our resources.

Often in science fiction of the pulp era, this included a strong urge to run off with "our women," reflecting both the gender attitudes of the time and the fact that the alien invaders were really little more than a Viking invader from a previous century, dressed up in a different costume, but still out to burn, rape, and pillage. (And possibly also reflecting the appeal to the typical readers of the pulp magazines of a cover image showing a scantily clad woman being attacked by an alien.) One way that this science fiction trope of the alien marauder putting human life at risk has strongly leaked out into the real world of science was in the warning of the danger we face from aliens that came from the physicist Stephen Hawking.

In 2010, Hawking raised a few eyebrows by suggesting in a documentary for the Discovery Channel that any alien visitors could simply be interested in making use of our resources, commenting: "If aliens visit us, the outcome would be much as when Columbus landed in America, which didn't turn out well for the Native Americans." As we will see, just as portrayed in some science fiction, we have actively sent messages out into space in the hope they will be intercepted by aliens, but Hawking suggested that instead of sending out a welcome, we should be attempting to conceal ourselves as much as possible to minimize the chances of contact. "We only have to look at ourselves to see how intelligent life might develop into something we wouldn't want to meet," he suggests.

In some ways, the warning comes a little late. The four current probes that will eventually exit the solar system all carry some kind of documentation of our existence. *Pioneer 10* and *11* had a gold-plated plaque showing our location in the solar system, the appearance of a naked man and woman, and an attempt to locate our star using pulsars; while *Voyager 1* and 2—reflecting a technology that has ironically all but disappeared from use on Earth—carried gold-plated metal discs, inscribed on one side with illustrations that are supposed to indicate how to play it, and the location of the Sun indicated by fourteen easily observed pulsars, and on the other with a groove like that on a vinyl record but including video images as well as audio. (The discs did not have the images of naked people featured on the earlier probes after complaints were received from the public.)

Given the size of the galaxy, the probes have far less chance of being picked up than a classic message in a bottle. Instead, if aliens are to detect our presence, it is far more likely to be thanks to various forms of light we send out into the universe. For over a hundred years we have been producing radio signals, initially few and feeble, but now in huge quantities. Some of these will head out into space, producing a sphere of radio influence around the Earth that in principle now extends over a hundred light-years. It's more than a little scary to think that aliens may base their understanding of Earth society on long-forgotten episodes of *I Love Lucy*.

In practice, the strength of radio signals drops off with the square of distance, so even our modern outpouring of radio will be difficult to pick up at any great distance. Two other possible techniques aliens could use once they spotted that the Sun had planets would be to compare the day and nighttime brightness of the planet—less differentiated when there is large amount of artificial light used—and to use the spectroscopic analysis of light from the Sun that has passed through our atmosphere to deduce its chemical composition and likelihood that this was consistent with an advanced civilization.

We certainly can't truly hide, as Stephen Hawking seems to want us to, but equally we are a small planet in a not particularly distinguished part of the Milky Way—so any aliens on the lookout for other inhabited planets would probably need to be fairly lucky to spot us. Which begs the question of what is happening in the field of flying saucers and alien visits. There is a powerful link between science fiction and apparent reality here, because a lot of reports of aliens and their craft tend to follow trends in how extraterrestrial visitors are described in fiction. Aliens were often referred to as "little green men" before the classic "gray" form with big heads and huge eyes become popular in science fiction—reported sightings followed suit. Similarly when the term "flying saucer" was first used, it was meant to describe how the craft moved (the way a saucer moves when it's skipped across water), rather than the shape of the ship. All the sightings since of saucer-shaped spacecraft seem to have been imaginings based on this misunderstanding and on common science fiction portrayals.

Because of the sheer number of stars in the universe it would be something of a surprise if we were the only intelligent species. There are billions of galaxies, many of them containing billions of stars. It seems very likely that a universe on the scale we know ours to be should have many planets supporting life, some of which, we would expect, would have civilizations more technologically advanced than we are.

There have been attempts to calculate the likelihood of alien life we could communicate with existing, notably the Drake equation, produced in 1961 by then National Radio Astronomy astrophysicist Frank Drake. It calculates the number of civilizations with which radio communication could be possible by stringing together a whole host of factors, from reasonably easy to estimate figures like "the fraction of . . . stars with planetary systems" (though even this figure has varied considerably over time) to pure guesswork values like "the fraction of life-bearing planets on which intelligent life emerges" and "the length of time such civilizations [with detectable technology] release detectable signals into space."

The equation has rightly been criticized for essentially compounding a whole string of made up numbers to produce an even more fictional value—this is arguably no more science than the tricked up formulae that PR companies produce to show which day of the year is least happy or to describe how to make the perfect sandwich. Others argue it is still valuable because it is a useful exercise in thinking about the key factors involved—but I'm afraid I'm firmly in the first camp. The original implementation of the equation produced answers in the range of 20 to 50,000,000. A more realistic range might be between 0 and 1,000,000,000. In effect, the true answer is: "We have no way of telling."

Theories for the lack of contact with aliens break down into three broad categories—that the aliens aren't there, that they haven't found us, or that they have found us but choose not to be seen. Each of these ideas has its appeal. It is certainly true that the circumstances for life on Earth are quite specific and it may just be that there are relatively few planets where life has developed beyond the scale of bacteria—perhaps just a handful in our galaxy. (The scale of space is such that it's quite possible for other galaxies to be teeming with life without anyone ever reaching us in the Milky Way). But scientists are wary of anything that suggests a special place for the Earth, when there is no good reason for it to receive such special treatment. Of course, if there were only one planet with life, then it would be bound to be, by definition, the planet on which the inhabitants were thinking "Why us?"—a variant on the so-called anthropic principle—so strictly speaking the "no special place" argument has no good scientific basis, but it is still hard to believe that we are unique.

The size of the Milky Way galaxy, let alone the whole universe, suggests that the second option that they are out there but haven't found us is eminently possible. It could be that there is plenty of intelligent life in the galaxy, but there is so much space to explore that with the fundamental physical limit of not being able to travel faster than light, the aliens just haven't arrived yet—and may never do so in the window of opportunity of the existence of intelligent life on Earth. After all, there are plenty of parts of the oceans on the Earth that we haven't got around to visiting, and by comparison with space, the oceans are pathetically tiny. Perhaps the majority of space will *never* be visited by living beings. The only problem with this explanation is that in such a scenario, it is still possible that some alien races could have managed to build self-replicating probes, which vastly expands their ability to explore.

Such interstellar exploration drones, like the nanotechnology robots in chapter 5, would be capable of duplicating themselves from the raw materials they find around them. A probe would travel to a planet, duplicate itself, perhaps many times, then the new probes would fly off to more planets, spreading through a galaxy well below the speed of light, but still using the power of doubling to quickly be able to take in more and more of the galaxy. And being unmanned, the probes would never have to return, simply sending back information to the home world and cutting down the potential journey times.

To have spread well across the galaxy, probes of this kind would have to have started out many thousands of years ago, and it's entirely possible that if one did visit the Earth, it wouldn't arrive during the few thousand years so far when human beings would have been able to record its visit. Such devices, sometimes called von Neumann probes after the mathematician John von Neumann who worked on the concept, do also have a practical problem that to make an easily replicable device mechanically would require it to be simple—and yet the task it has to perform, refining ores, producing complex machinery and electronics, is very complex. We simply couldn't build a device that could replicate itself from raw materials like this at all, let alone one that then had the ability to power itself out of the Earth's gravity well and navigate between the stars. It may well never be possible.

The final option for consideration is that aliens know that we're here, but don't want us to be aware of their existence. Perhaps they have cloaking technology (see page 225) to be able to move among us unseen. Perhaps they have some kind of noninterference directive as in *Star Trek* (though that never seemed to stop Captain Kirk getting involved) or they might simply regard us as too unpleasant or inferior to bother to mix with. If either of the second or third cases holds true, we could still at some point still encounter aliens—and it's possible that like the aliens of so many B movies, and those predicted by Stephen Hawking, they will prove unfriendly and will want to exterminate us. Equally, they could have friendly intentions. But in either case, given the lack of proof to date, I wouldn't bother to take out insurance against the possibilities of an alien invasion.

So if we don't have any evidence of aliens falling in with science fiction and giving us a visit, could they at least be trying to signal to us? Messages from the great beyond have an interesting history in science fiction. Arguably one of the first was from the great engineer Nikola Tesla, who might not have been intending to create fiction when he speculated that a regular electromagnetic signal he was picking up could be from aliens, though the chances are it was produced by a simple natural cause, like radio emissions produced by electromagnetic activity on the outer planets. However, two alien messages stand out in fiction—one friendly, the other far less so.

The first came from the science fiction of a working scientist. Occasionally scientists try their hand at fiction—often substituting strong science for anything approximating to good storytelling, but the British astrophysicist Fred Hoyle managed at least one approachable story. Hoyle was something of a maverick in his field, dreaming up with a couple of colleagues the "steady state" theory for the origin of the universe, which for a while had more credibility than the big bang, until new data made it unlikely. (Hoyle never let the theory go, pointing out that if it had been patched up to match observations as much as the big bang had, it too could still be applied.) Fred Hoyle was also responsible for giving support to the controversial idea of panspermia, which suggested that life on Earth originated from an extraterrestrial source. Yet his best work on the origin of elements in stars and supernova was a superb contribution to astrophysics.

Some of Hoyle's novels were infamously esoteric, but *A for Andromeda,* perhaps because it was originally developed for television in 1961, was more of an easy read and told of the dangers of following the instructions in an intergalactic message. In the book (and TV show), a new radio telescope picks up a signal that seems to originate from the Andromeda Galaxy. The complex data in the signal proves to be a computer program, which gives instructions for building a computer unlike anything ever seen. This machine then collaborates in growing a monstrous living organism—but when a female scientist is somehow drawn to take hold of a pair of terminals on the computer she is killed. In the process the computer seems to have analyzed her structure and grows a new organism that is a clone of the dead woman and that seems to be the beginnings of an alien invasion force.

Apart from the cleverness of making the final version of the alien an attractive human being, this story line does get over that huge problem of the time taken for interstellar (and in this case, intergalactic) missions. The Andromeda Galaxy is 2.5 million light-years away—it takes light 2.5 million years to reach us, and the alien race that created the signal in the story could well have long been dead when the action took place. But by merely transmitting information and creating the physical entity on the spot, it was possible at least to make that phenomenal journey and arguably perpetuate their species. Unfortunately for the people of Earth, this alien certainly did not come in peace.

Overcoming the impossibly large distances of space was also the premise of the second fictional alien message, though in this case the aliens were altogether more friendly. *Contact,* originally a 1985 novel by science communicator Carl Sagan, was made into a movie starring Jodie Foster that is often considered one of the best sci-fi movies for its science content and portrayal of a scientist. In the movie, an alien signal carries not the seeds of an invasion force but the instructions to enable human beings to join the star-traveling civilizations.

The script centers on Dr. Ellie Arroway, who is employed by SETI, the Search for ExtraTerrestrial Intelligence program, at the Arecibo radio telescope in Puerto Rico. (Interestingly, in real life we have sent out a burst of information from Arecibo, our single largest dish, in an attempt to communicate with others in the universe.) Arroway and her colleagues receive a signal from Vega, a star also known as Alpha Lyrae, which is the fifth brightest star as we observe it, about 25 light-years in distance. The signal incorporates video of Adolf Hitler's address to the Summer Olympics in Berlin in 1936—a signal that would have had time to have reached Vega and be broadcast back as acknowledgment of our existence.

Somewhat like Hoyle's story, the signal also contains the information required to build an incomprehensible alien piece of technology, which in this case is somehow capable of transporting Arroway to a planet around Vega (Sagan employs a series of wormholes for this journey). She makes the trip briefly, though as far as observers from Earth were concerned it appears that the device sent her nowhere, letting her fall through into a safety net, so we are left in doubt as to whether the trip was real or in Arroway's imagination.

In reality we have received no plans for a hi-tech machine, and traveling to Vega (which may have a planetary system, but so far has no candidates capable of supporting life identified) is not on the cards, especially through wormholes. But we have received a pair of truly fascinating signals from space that could be considered our *A for Andromeda* moments to date.

The first signal was detected by PhD student Jocelyn Bell, working for radio astronomer Antony Hewish at Cambridge in July 1968. This was a high-speed pulsing of a radio frequency—no information, yet the broadcast had the clockwork regularity of a mechanical device. Could aliens be signaling to us? Bell and Hewish claim not to have thought so, despite jokingly labeling the signal "LGM-1"—standing for Little Green Men, a common tongue-in-cheek designation for aliens in the 1960s. They said that they never took this possibility seriously, though this was a time when the very mention of alien life would have been frowned on in fusty academic circles, so even if they had suspicions it is unlikely that they would have admitted it.

When I was at Cambridge studying physics, I spent some time at the radio astronomy laboratory, building a basic radio telescope with a couple of my fellow students as a final year project. It was enjoyable work, but also a very isolated experience. The observatory is sited well away from the city center to limit local radio pollution, and we were often working outdoors with surveying equipment and crude wire frames like large bedsteads, or crouched in a freezing wooden hut poring over the output of a teleprinter. It would be easy to imagine the young Bell, perhaps on her own long into the night in this spooky, isolated site, when this unprecedented signal was picked up. Did she really not give a moment's thought that something, something intelligent, was out there, signaling to her?

It turned out that despite its apparent mechanical regularity, the signal was natural, the first evidence to come from radio astronomy of the existence of pulsars. Pulsars are very dense stars—neutron stars—that rotate at immense speeds, sending out pulses like a radio wave version of a lighthouse. These pulses can come every few seconds or as frequently as a couple of thousandths of a second apart, so fast does the star rotate. The result is, indeed, a signal that is quite unnervingly artificial sounding.

The LGM-1 source was relatively quickly explained, but a second signal, the so-called "Wow! Signal," has never had the same certainty about its cause. This was the best result yet to be discovered by the real SETI on which Ellie Arroway's work in *Contact* was based. SETI emerged from Project Ozma, set up in 1960 by Frank Drake (see page 127) to examine radio output from Tau Ceti (then thought to be one of the most likely stars to have life-supporting planets) and Epsilon Eridani (one of the closest stars to us) to look for signals. Drake had to face up to what is arguably the biggest challenge of would-be signal finders of knowing where to look and what to look for.

The problem is that we can't scan every frequency for anything unexpected, but need to home in on something that another intelligent life-form is likely to use, either for their own communication purposes, or in a deliberate attempt to communicate with aliens like us. Drake chose the latter option and thought that an obvious choice would be to use frequencies near the spectral lines for hydrogen and for hydroxyl groups, the components of water, generally thought to be likely to be an essential for life. It's a big leap of faith. This is reasoning that seems sensible to us but would not necessarily be meaningful to an alien—but Drake had to look somewhere, and so Ozma concentrated on this so-called "water hole" section of the waveband. Sadly, nothing interesting was detected.

SETI has continued on and off ever since, always struggling to be funded as its science fiction feel made mainstream scientists wary of being associated with it. From the 1970s to the 1990s, SETI projects had some U.S. government funding, notably through NASA, but since 1995, any progress has relied on private donations as the government support was withdrawn. It was as far back as 1977, at the Ohio State University's "Big Ear" radio telescope, that a volunteer observer, Jerry Ehman, noted something strange—SETI's only real success.

One thing those working with radio telescopes have to get used to quickly—and this was especially true in the early days—is interpreting a mass of data on obscure-looking printouts. The Big Ear produced sheets of paper with a series of numbers, denoting the strength of the signal being received. The code used showed a blank for the weakest intensity, going up through the numbers and then the letters, so that an A was the equivalent of strength 10 and U, the highest value measured, was strength 30. Generally speaking, the signals detected were in the range between blank and 3. But August 15, 1977, Ehman spotted a sequence of characters that burst out of this background, easily visible to even the untutored eye. He circled them in red pen and wrote "Wow!" on the sheet to highlight their unusual nature.

The characters read 6EQUJ5—a burst of radio waves that rapidly rose in intensity, peaking at an extremely strong signal. What's more, the duration of this signal was seventy-two seconds, exactly the period expected for an extraterrestrial source as the Big Ear's window of view scanned across the sky via the rotation of the Earth (the telescope itself was fixed). The source appeared to be a location in the Sagittarius constellation, though it could not be pinned down to a specific star. The staff eagerly awaited a repeat—but none ever came.

Since the detection, a number of explanations have been considered, notably an earthbound source and a one-off natural event in space, but the signal remains unusual in the annals of the SETI search. When the Wow! Signal was detected observations were still being made on a limited range of frequencies, in this case close to the hydrogen spectral line, but later work used devices capable of working across many millions of channels with no further success to date. The search continues.

Originally, of course, our view of alien life was much more limited. We assumed that we would find life on the nearby planets once we realized that they were, in effect, other Earths, and science fiction followed this lead. Due to a confusion over the translation of a word meaning "channels" as "canals" and some epic self-deception, it was thought for several decades that Mars was, or had been at one point, home to intelligent life that had suffered as the conditions on the planet deteriorated. Meanwhile Venus, the planet closest to Earth in size and permanently shrouded in clouds, was thought to be a tropical paradise under its permanent clouds.

In reality it has turned out that neither Martians nor Venusians are going to be visiting us any time soon. Mars does seem to have had running water at some time in the past, but as yet we have found no evidence of life, or even the former existence of life, even down at the microbial level. And rather than a humid paradise, Venus has turned out to be a fair parallel to hell with surface temperatures averaging around 460 degrees Celsius (860 degrees Fahreneit) and reaching as high as 600 degrees Celsius (1,100 degrees Fahrenheit). This is significantly hotter than the surface of the Sun-hugging planet Mercury. The metal lead would run liquid on the surface of Venus.

What's more, although Venus does have a thick atmosphere, one of the reasons it was thought to be a hopeful home for life—it is nothing like our atmosphere. Venus is swathed in an extremely dense layer of carbon dioxide, giving it an immense atmospheric pressure at the planet's surface. It is over ninety times the pressure on Earth, meaning that any probes we have sent have failed within minutes under the assault of the Venusian weather. The over-the-top greenhouse effect that all the Venusian carbon dioxide produces is responsible for those incredibly high temperatures. Throw in sulfuric acid rain, and it is obvious that this is an unlikely home for anything living that bore any resemblance to life on Earth.

Surprisingly, the best hope for finding some very low level of life is one of the moons of Jupiter, a location that would once have been considered far too cold for anything to live. All but Io of Jupiter's big four moons are thought to have oceans of water hidden beneath their surfaces. Despite Jupiter being so far from the Sun, these lunar oceans are likely to be kept warm enough to be liquid by the tidal energy from the planet. The most likely candidate for hosting life is Europa, whose unusually smooth surface is an ice layer, which is very likely to cover a liquid ocean, while any water on Ganymede and Callisto would have to be much deeper beneath a rocky surface.

As yet, then, we could still be alone in the universe, and science fiction has to be our main source of contact with imagined alien life, whether it is out to befriend us or to bring an end to life on Earth. But one thing is certain. Science fiction writers love finding ways to put the human race in peril.

10.

END OF THE WORLD, PART THREE

It would seem very difficult for young people today to comprehend, but when I was a teenager, the end of the world at our own hands was an everyday consideration. Back in the 1970s we genuinely felt that it was likely there would be a nuclear attack as an escalation of the Cold War and that life as we knew it would come to an end. I did my masters at a university that was often under the flight path of military aircraft, and I remember regularly feelingly nervous at the sound of those jets flying low overhead. Atomic Armageddon was a reality for us, not implausible fiction. In a previous book I quoted writer Sue Guiney, describing how this impacted on her as a child in Farmingdale, New York:

> [The Cold War] was the backdrop of my childhood. All the fears, real and imagined, are still there inside me, almost on a cellular level . . . As a seven-year-old, I remember having air raid drills where we were lined up in the school hallway and told to sit curled up facing the walls. I remember thinking that my small curved back would make a perfect target for a falling bomb, and I had nightmares about it for years.

The interesting thing about disasters and science fiction, which is why this chapter is going to be relatively short, is that unlike most sci-fi, disaster science fiction has tended to be far more reactive to the real world than innovative in the ways it has suggested that life on Earth could be put in danger. With the exception of the kind of alien invasion described in the previous chapter, the ways the world can end have not tended to be new possibilities, inspired by science and technology, but have played out the disaster that was currently most in our minds as accurately as possible.

So, for many years after the atomic bombs were dropped on Hiroshima and Nagasaki at the end of the Second World War, the dominant fictional mechanism for life as we know it to come to an end was an atomic holocaust. Some science fiction took us through the bleak and hopeless experience itself, particularly in the 1960s, when the portrayal of the futility of life was a central theme for the so-called "new wave" science fiction writers. More portrayed a post-nuclear world and humanity's attempts to pick up the pieces.

Sometimes these dreamed-up futures could lead to unlikely and novel possibilities, as in John Wyndham's 1955 novel *The Chrysalids,* where mutants caused by nuclear fallout begin to transform the nature of the human race, whether it is in simple physical deformation or strange mental abilities. Equally the post-apocalyptic world could hold out a kind of romantic hope, as seen most clearly in the 1959 novel by Walter M. Miller Jr., generally considered a classic of science fiction, *A Canticle for Leibowitz,* in which civilization is rebuilt, in part due to the initially ignorant preservation of technology by Catholic monks.

Other authors have brought civilization to an end through disease, often a result of genetic modification or from misguided attempts to produce biological weapons. Alternatively, the cause of our near extinction could be extraterrestrial without the need for implausible aliens. Reflecting the probable destruction of the dinosaurs around 65 million years ago by the impact of a huge meteorite, it has not been uncommon for debris from space—or radiation from a misbehaving Sun or more distant gamma ray bursts that have put our survival on the line. At the time of writing, the disaster du jour is almost certainly climate change.

There may be disagreement in right-wing political parties, but for most who are prepared to listen to the science, pretty much the biggest threat that we face as a race at the moment is climate change. In earlier science fiction, climate disasters tended to involve a return of an ice age. (And it has even crept into modern disaster movies in the form of *The Day After Tomorrow*.) We know that humanity has nearly been wiped out several times in the past as ice has crept down over the continents, and another ice age has presented a credible threat, whether as a natural resumption of the current ice age (we are just in a period of interglacial warming at the moment) or some kind of nuclear winter. But now, of course, the big chill has largely been replaced by global warming.

So common is science fiction with a climate change theme at the moment, particularly for the young adult market, that it has been given a subcategory of its own, known as cli-fi. There is plenty to work on as a result of the dire warnings of the climate scientists. Take sea-level rise. If the entire Greenland ice sheet were to melt and end up in the ocean, it would raise sea levels by 23 feet. This would have a catastrophic impact on cities near to sea level like New York. And if it were followed by the Antarctic ice cap melting (admittedly unlikely, but we are talking fiction), the rise could be another 200 feet—and that's without storm surges.

Climate change would have a massive impact on everything from agriculture to the availability of water. At the same time as huge swathes of low-lying countries like Bangladesh disappearing under water, others would be in a state of permanent drought. Climate change rarely makes for enjoyable reading, but it has fostered many a disaster novel.

All these require very little deviation from the present reality. They are disasters we could easily envisage or make with our current science and technology. It is only in a small number of cases that science fiction has come up with a truly original way to end the world as we know it. We might see the world taken over by intelligent computers, as we will see in chapter 17, but perhaps the best opportunity comes from nanotechnology. As we saw in chapter 5, there has already been a portrayal of destruction by rampant nanobots, tiny robotic creatures that can assemble (or disassemble) molecules to construct new matter and so could eat their way through anything and everything.

Arguably nanotechnology still offers the most new opportunities for ending the world with a true science fiction twist. In my own young adult novel, *Xenostorm: Rising*, a small band of humans have tiny nanobots in their body that maintain them, keeping them alive forever. They have been attempting for many years to produce a child in which the internal nanobots become a sentient hive being, working with their host to form a superhuman outcome. This happens to the main character of the book, Davy, who discovers that an earlier attempt to make this happen has produced "the abomination."

This is a being constructed only of nanobots, whose human flesh has been eaten away. It is capable of absorbing others with nanobots, merging their consciousness into a single entity. It discovers there is a mechanism by which it could blast apart Davy's sentient nanobots and spread them to the world. This would give indefinitely long lives to everyone in the world—but would also give the abomination a chance to absorb the whole population, turning the entire human race into a single super entity.

The themes I have used reflect how much is open still in the use of nanotechnology both to examine the "What if?" of an end of the world as we know it caused by the nanobots, and to consider what a world where no one died, other than by accident or murder, would be like. That too ends the world as we know it, but in a very different way. The main character, Davy, must decide if he has the right to impose unending lives on the world. It seems likely because of the range of possibilities available that we will see a lot more nanotechnology driven science fiction stories in the future.

A final possible way to end civilization is arguably to just carry on the way things are at the moment. This is often seen as the way that climate change will overwhelm us, but many argue that the same laissez-faire attitude puts humanity in danger of subservience to corporations who may make token claims to have the good of humanity at heart (think of Google's former corporate motto "Don't be evil"), but in reality are driven by hard cash and shareholder returns, trampling individuals under foot.

We started this chapter with the chill of the cold war threat of atomic annihilation. And during the 1960s and 1970s the world became familiar with antinuclear demonstrations. But the protests weren't limited to calls for nuclear disarmament. They also targeted nuclear power plants. Whether inspired by the power stations' initial association with the bomb program, or later fears after accidents like Three Mile Island and Chernobyl, green organizations have mostly campaigned against nuclear power, despite it being a practical way to generate electricity with negligible contributions to climate change. Finding suitable sources of energy in large quantities has become an essential for modern life, a need that has long been reflected in science fiction.

II.

ATOM WORLD

Like most people, I have had little direct experience of power production. I've never been to a power station and I tend not to think about where the power from the sockets originates, except in very broad terms when considering the impact of fossil fuels on global warming, or the alternatives for green generation of energy. Yet I had an underlying belief about the future when I was younger that I got from science fiction, which was that energy would become more readily and more cheaply available with time. Whether it was Isaac Asimov promising walnut-sized nuclear generators carried on our persons or descriptions of worlds where power seemed to be as free and easy as sunlight, there was little sense in science fiction that availability of energy (or the impact of its generation on our environment) would ever be an issue.

Probably the biggest gap of all the topics between what was envisioned in fiction and the reality we have experienced is in our attitude to energy. If we had the kind of near-infinite supply of (almost) free energy often promised, the world would be very different. Getting into space would be trivial. Water shortages and drought would disappear, because the Earth has vast quantities of water—it's just a matter of getting the salt out of it and getting it to the right places . . . which takes a lot of energy. Comic book science fiction, particularly, has always played fast and loose with energy. Watch any superhero in action and you will inevitably see numerous problems because the conservation of energy goes out of the window. And the early writers of pulp science fiction added to the misuse by merrily producing energy out of nowhere.

Even our old friend H. G. Wells was guilty of this sin against one of the most fundamental laws of physics, the first law of thermodynamics. His *The First Men in the Moon* used a "magic" source of energy in cavorite, an artificial substance that was supposed to block the influence of gravity, enabling a spacecraft to float away from the Earth. Leaving aside the difficulties that arise from the thought of blocking gravity, if such a substance existed, using it to fly into space would be a trivial application, as properly applied, it would provide that free and near-infinite source of energy. All Wells's Mr. Cavor needed to do was to paint the bottom of each paddle in a waterwheel (for example) to produce a machine that generated energy from nothing—a perpetual motion machine. When the cavorite sides were down the paddles would be weightless, while the paddles on the other side with the metal side down would feel the pull of gravity. Result: the wheel would turn itself.

The whole concept of antigravity cropped up frequently in early science fiction. The writers were not thinking through the energy-from-nowhere implications of the idea, instead making a reality of a feature of some human dreams—the ability to float through the air. Not to fly with effort like a winged bird, or to glide, or even to travel under power as discussed in the Iron Man chapter (see page 87), but simply to float as if gravity had no effect. This was fantasy in a science fiction guise. The closest that science has come to this is various forms of levitation, which rely on opposing the relatively weak gravitational force with electromagnetism.

This effect is impressively demonstrated by superconductors, which have the property of expelling magnetic fields—so a magnet placed over a superconductor will float mysteriously in the air. And, in our most science fiction–like mode of transport, those same superconducting magnets are used to float maglev (magnetic levitation) trains above their rails, enabling the experimental MLX01 to smash the rail speed limit, achieving 361 miles per hour as it hovers a tiny distance above the track. Powerful magnets have even levitated a frog with no special harness, leaving the poor animal floating in midair as the magnetic field aligns the water molecules within its body to act as tiny internal magnets, producing a repulsion that is enough to support the frog's weight. But all these applications fail as an electromagnetic equivalent of floating under antigravity because they rely on a source of electromagnetism outside the vehicle. They aren't self-contained.

There is one possibility, which is to use an inverted version of the frog demonstration. Here a craft would induce such a strong field in the ground or in the air below it so that the vehicle would float. But this would require a massive power source in the vessel (and could have some disastrous effects on anything it flew over). All the evidence is that such an attempt at antigravity is unlikely to be successful—and is certainly no source of power, but rather a huge consumer of energy, which would have to be stored on board or routed to it somehow. It's interesting that by the 1970s, with, perhaps, a wider understanding of the nature of gravity as general relativity was demystified, science fiction pretty well gave up using antigravity as a device. The ability to float through the air returned to fantasy, in the likes of Superman, where it belonged.

The energy that we use on Earth outside of science fiction has always primarily come from the Sun. This heats our planet and powers the weather system, so can be passed on both as wind and as water, when the system evaporates water from the sea, transports it in the atmosphere, and dumps the water on a high piece of ground like a mountain, giving the high-altitude water potential energy that can be tapped in dams and waterwheels. A more indirect route for the Sun's energy to be made available is as chemical energy. Here it is typically stored in plants by photosynthesis and transferred to an animal by eating before being released in animal muscles. Or we might take the more direct and dramatic chemical energy release of burning a substance that has stored up solar energy to produce heat.

Early science fiction realized that something more was needed than our traditional Sun-based sources to power the energy-guzzling devices beloved of the science fiction fraternity. By the time E. E. "Doc" Smith began writing in 1915, the prolific writer of space operas, whose later Lensman series, as we have seen, was one of my first introductions to classic pulp science fiction, it was commonplace to come up with a magical source of energy. In Smith's earlier Skylark series, there are ships that are powered by a mysterious "metal X" that provides "infra-atomic energy" producing instant motion. Magic with a scientific wrapper.

Another early example of a mysterious energy source was the "Vril" dreamed up by Edward Bulwer-Lytton. These days, Bulwer-Lytton is probably best remembered for writing what is often described as the worst-ever opening line of a novel (since appropriated with much relish by Snoopy in the *Peanuts* cartoons): "It was a dark and stormy night." In his less celebrated (or mocked) book *The Coming Race,* Lytton described what seems to be a distant relative of electricity, described as "atmospheric magnetism," controlled by staffs of power that seem to belong more to a Tolkien-style epic than science fiction. Vril had no basis in reality, but it proved a very popular concept, widely adopted in product names of the time, just as radium would be later. If your product had a name with "Vril" in it, it seemed to suggest the product would give the purchaser boundless energy. In the UK there is still a beef extract used to make hot drinks with the name Bovril, though few of the (mostly elderly) purchasers realize that they are buying into a Victorian dream of energetic release.

It wasn't long before most science fiction writers were happy to follow the early lead of the H. G. Wells book *The World Set Free* and make their "magic" source of energy nuclear power, a source that could provide both a powerful ally and a terrifying enemy. Back when *The Space Merchants* (see page 37) was written in the 1950s, the biggest obsession of science fiction writers was with anything that could be labeled atomic, hardly surprising given the impact of the nuclear bombs at the end of the Second World War and the newly developed idea of electricity generated by the power of the atom. The term "atomic" appeared everywhere, even as a label for chili fries in diners to indicate they were spiced up and energetic. Whether it was the giant ants of the movie *Them!,* the protagonist in *The Incredible Shrinking Man,* or the radioactive spider bite in the Spider-Man story, nuclear radiation was portrayed as being responsible for many ills, oddities, and marvelous transformations of our world.

Because of politics and fear (and our frustrating lack of ability to create superheroes), the real nuclear world has been very different—but we are at long last getting close to possibility of generating energy from nuclear fusion, the power source of the Sun, and take radiotherapy for granted as part of our medical kit bag. Unlike Dr. Strangelove, we might not have learned to love the bomb, but we have an uneasy truce with nuclear power. But it is a fragile truce, as demonstrated by Germany's withdrawal from electricity generation by nuclear power after the disaster at the Fukushima plant caused by the tsunami in 2011. Even though that accident happened under conditions that could never occur in Germany, and resulted in no deaths, the Germans were prepared to abandon our best current means of generating energy without greenhouse gas emissions.

While we certainly are unlikely to ever have power from nuclear that is too cheap to meter, as was once assumed, we currently have sufficient power in developed countries to meet most of our needs, and apart from avoiding emissions, the biggest issue we face, something that again tends to be glossed over with magic solutions in science fiction, is getting power easily from A to B. At the moment we use high-voltage AC power lines, and though there is now talk of moving to high-voltage DC, which has less power loss when carrying large amounts of power over long distances, what we don't see is science fiction's favorite means of moving power, the wireless transmission of energy.

Of course this is entirely possible—it's how we survive. After all, as we discovered at the start of the chapter, most of our energy comes from the Sun directly or indirectly. (The alternatives are tidal energy, primarily from the gravitational effects of the Moon, and geothermal energy, in part produced by the Earth's internal heat.) And there are no cables snaking through space. We receive vast quantities of energy from the Sun wirelessly, just in the form of light. Every photon of light is a tiny packet of energy and the Sun has plenty on offer.

The Sun's output is around 400 billion billion megawatts, of which around 89 billion megawatts arrive at the Earth. This may only be a small fraction, but it is around 5,000 times global energy consumption. This is impressive, but unless you have the sheer capacity of a star, using light isn't a great way to transfer energy from place to place. Firstly we need to do something practical with it when it arrives. The ways we have of converting light to electrical energy are fairly inefficient. While specialist solar cells have achieved between 25 and 45 percent efficiency in the lab (and even then, more than half the energy is being lost), typical efficiencies are more likely to be in the 10 to 20 percent range for conventional cells and less than 5 percent for thin film cells. Compare this with a typical 93 to 95 percent efficiency for power lines as a means of transmission.

The other problem with light as an energy transmitter is that cramming a whole lot of energy into a beam, whether visible or, say, in the microwave region, makes it potentially very dangerous for anything living that gets in its way. You can't just send the beam through open air without risking something intercepting it and being destroyed—yet if the beam has to be sent through some kind of protective structure, the infrastructure costs and inconvenience make power lines seem attractive again.

A number of inventors in the past have dreamed of wireless power transmission, notably Nikola Tesla (see page 106). Tesla built an experimental station called Wardenclyffe with a huge transmitting tower to experiment with power transmission. Not fully understanding the nature of electromagnetism, Tesla thought it should be possible to send power as a standing wave in the Earth, with some kind of electromagnetic return circuit through the air. His devices, which took a huge amount of power, were certainly capable of inducing a current remotely, but this was nothing more than electromagnetic induction, the same process that is used in electrical transformers.

The changing electrical current in a wire produces a magnetic field, which generates a current in a separate, unconnected wire. But unlike Tesla's imaginary power transmission science, which was supposed to work at any distance without dropping off of power, induction drops off very quickly with distance, making it unsuitable as a means to transmit power. To make matters worse, it is very nonselective, producing electrical current wherever it can, resulting in all kinds of unwanted electrical effects in nearby wires and metal constructions. You can see induction in action if you take a long fluorescent tube and place one end on the ground near to a high-voltage overhead power line. The tube will light up with no wiring as a voltage difference is induced between the earthed end and the end nearer the power line. But this effect is generally without value for practical transmission of power.

Whichever way the power is moved from place to place, there is one power source, or more accurately a power storage mechanism, that beats nuclear power, and that is antimatter, famously the fuel of the USS *Enterprise*. Antimatter engines sound like science fiction, and the mechanism used in *Star Trek* is entirely fictional, but antimatter itself is real enough. Antimatter is the same as ordinary matter, but the particles that make it up have the opposite electrical charge to those in everyday atoms.

Where, for example, an electron has a negative charge, the antimatter equivalent, the antielectron (better known as a positron), has a positive charge. When matching matter and antimatter particles are brought together they are attracted by their opposite charges, smash into each other, and are destroyed. The particles' mass is entirely converted to energy, and while individual particles have very low mass, Einstein's famous equation $E=mc^2$ tells us that the energy produced will be equal to the mass of the particles multiplied by the square of the speed of light. That's a big number.

A kilogram each (around 2.2 pounds) of matter and antimatter would generate the equivalent of a power station running for over ten years. Antimatter is the ideal source of energy for a space vessel—though it should be emphasized that it isn't, as some people think a "clean source of energy" because to produce antimatter takes more energy than it releases and that energy has to be produced by conventional means, which can be as green or as dirty as your generation method allows. Antimatter itself merely provides a very compact way to store that energy and to be able to release it at will—useful if you are building the classic science fiction spaceship. Just how compact it is can be seen by comparing antimatter with some of our more familiar sources of power, seeing how much energy we can get out per kilogram of fuel available.

Gasoline and natural gas, the fuels that most of us have direct experience with in everyday life, are surprisingly efficient at packing in the energy. A kilo of gasoline has around 15 times the energy of the same mass of the explosive TNT. (The only reason TNT seems to be more energy-packed is that it burns much quicker, producing an explosion.) And gasoline also beats the best current batteries by a factor of 100. Natural gas is even better than its liquid equivalent, holding around 1.3 times as much energy per unit mass, though it tends to take up more room.

But no one is going to power a spaceship by fueling it up with gasoline at the nearest service station. Hydrogen, the power source used by the Apollo rockets, beats gas by a factor of 2.6, our best practical source to date. Yet a nuclear fission engine of the future could, in principle, pack 2 million times the energy into the same mass. Make that nuclear fusion instead and we're talking 6 million times better. And this is where we can see just how efficient antimatter is as a fuel source. For each pound (or kilogram) of antimatter we get 2 *billion* times as much energy out as we would from the same mass of gasoline.

Despite these impressive numbers, we won't be using antimatter any time soon. A spaceship would require at least a few tons of this fuel to get up to the kind of speeds needed for any meaningful interstellar travel. And antimatter is not easy to make. It is perfectly possible. As is mentioned in Dan Brown's book *Angels & Demons,* antimatter is produced at the same CERN laboratory in Europe as hosts the Large Hadron Collider, though in a totally different experiment (about the only "fact" in Brown's book that is actually true). But production is slow. At the moment, the whole world's annual production of antimatter is less than a millionth of a gram, so we aren't going to be building antimatter spaceship filling stations any time soon.

As described in my book *Final Frontier,* plenty of effort has been put into the possibilities of enhancing current power sources for spacecraft. The hydrogen/oxygen rockets of today are indeed more efficient than using gasoline, but they can't compare with the possibilities of using nuclear fuel—that factor of a million or more makes a huge difference. For traveling around the solar system, nuclear power would be extremely valuable today by combining the kind of nuclear reactor used on a submarine with an electric-powered ion thruster that uses electrical power to fling charged particles from the rocket exhaust, but this has nowhere near enough power to get a ship heading for another star.

The best nuclear option is probably nuclear fusion—the power source of the Sun and hydrogen bombs, but the problem arises of how to practically make use of fusion as a means of propulsion. An apparently idiotic suggestion dating back to the 1940s that actually makes a fair amount of sense is to feed a series of small nuclear bombs out of the back of a ship that is protected by a "pusher plate." This coated, curved piece of steel absorbs the shock wave produced by the bomblets exploding, propelling the ship forward. As long as the bombs are small and exploded sufficiently far behind the ship, the result should be to push it forward without damaging the structure, while a shock absorber system would ensure that each of the many pushes does not produce so much acceleration that the crew are liquidized by the sudden pressure.

This design of a nuclear bomb-powered ship was even given a detailed workup between 1958 and 1965 in the remarkable Project Orion, a U.S. response to the shock news that the USSR had managed to get a satellite into orbit first. The idea was to leapfrog the Soviet show of strength with a ship that could get men to Mars by 1966. (Now that really would have been bringing science fiction alive.) NASA is well known for coming up with proof of concept designs that are pretty close to the ideas of fiction writers, but Orion, predating that agency, dwarfed even most of NASA's dreams. Where the space shuttles that have carried the biggest crews into space to date weighed 100 tons, Orion was expected to weigh around 10,000 tons.

Not only was Orion intended to cope with around 150 people on board, it was expected to manage a payload of thousands of tons, the kind of capacity needed for building vast space stations or taking on the colonization of another planet. It seems remarkable now, when nuclear explosions are banned even for test purposes, that the plan was to use a stream of fission bombs to get Orion off the ground, not just to power it once it was in space. Although, inevitably, Orion never flew, it was tested in principle by placing metal spheres near nuclear test bombs and observing their flight, and the idea was rekindled in the 1970s in the form of Project Daedalus. In this more modern design, the ship would sensibly start in orbit, to avoid any issues with nuclear explosions on Earth, and would use nuclear fusion instead of fission as a power source. As well as releasing more energy for the same amount of mass, always critical in a space vessel that has to carry its own fuel, fusion has the advantage of not being limited to a minimum critical mass, so it would be possible to use a stream of much smaller bomblets—up to around 250 a second—to propel a more manageable-sized craft.

Daedalus was never anything more than a thought experiment dreamed up by enthusiasts, but was designed as an unmanned probe that could reach the nearby Barnard's Star (at the time thought to have a solar system, though it has since proved not to have) in around fifty years, not a bad time to cover six light-years. Over the years since, there have been many attempts to come up with technology designs that would enable ships to reach the kind of speed that would make interstellar journeys possible. Often they take a basic drive and then add on extra means of acceleration, from deploying solar sails to giving the ship a boost with an electromagnetic catapult called a "mass driver." But no matter how good the engine there is inevitably a limit—the speed of light.

If a ship could get near to the speed of light, the effects of special relativity would mean that, where to those left behind on Earth the trip might take many decades, to the astronauts, the journey might only take a year or two, making interstellar journeys seem feasible. In practice, because it takes more and more energy to accelerate a tiny amount as a ship gets close to the speed of light, even the best possible technologies could only be expected to deliver a fraction of light speed—perhaps 10 to 20 percent. This makes a round trip to any star worth visiting impossible for humans unless they are in some form of suspended animation, a technology that has its own problems, and so thoughts turn back to science fiction and a faster-than-light drive.

Generally speaking, faster than light is considered to be one of SF's signature shorthand cheats. A dip into fantasy in order to get around insuperable problems and make stories work, but that has no solution in the real world. Science fiction can deploy ideas like "hyperspace," for example, a concept that has no real equivalent in physics, as a shortcut to get around the light speed barrier. The term itself predates science fiction, appearing as far back as 1867 in a mathematical treatise, simply used as a term for space that had more than three dimensions. But by the 1930s, early SF writers of the pulp era like John W. Campbell were using the term explicitly as a "get out of jail" card for the difficulty of crossing vast interstellar distances.

The details of flying through hyperspace were rarely explained as it rapidly became a convention that could be used by a writer, where the readers would understand what was meant without anyone needing to know how it worked. Broadly, the idea was that there were more than three spatial dimensions, with extra dimensions that we three-dimensional creatures never directly experienced. This is something that would later be echoed in the real world by string theory, the very speculative physics concept that requires there to be as many as nine spatial dimensions. But this doesn't provide a lifeline for bridging science fiction and the real world as, if those dimensions existed, they would be rolled up so small that we could observe them.

This makes string theory an unsuitable basis for hyperspace, because hyperspatial travel required that some of the extra dimensions, in some unspecified way, would act as a shortcut that would make it possible to fold space and get to a distant location without passing through the real-world space in-between. Tiny curled up dimensions wouldn't do the job. Even when SF techniques for faster than light travel make use of an actual aspect of physics—as in *Contact* (see page 131) where a wormhole or an Einstein-Rosen bridge was used—there are problems. The wormhole is a theoretical concept that links two points in spacetime because of a distortion in the curvature predicted by general relativity—but the realities of using any such technique are far from ideal.

Despite its attractions, a wormhole is not an everyday object. We don't know how to make one, or how to get it in the right place. If we could make one, theory suggests it would collapse the moment anything tried to pass through it. The only way of keeping it open would require negative energy. Admittedly negative energy has been observed, but the only known examples of it exist on a disappearingly small scale and couldn't be practically deployed to perform a task. This is a shame, as if negative energy could be managed, there is a small hope of something more practical than wormholes. Something that, conveniently for the theme of this book seems almost a perfect match for the engines of the USS *Enterprise*—a warp drive. And this idea has not come from a science fiction author, but from NASA.

That statement probably needs a little expansion. Mainstream NASA would raise a collective eyebrow at the thought of a warp drive, but the organization has always encouraged its employees to push the envelope, and one of them, Harold "Sonny" White, has done more pushing than most. Picking up on a concept dreamed up by physicist Miguel Alcubierre, working at the University of Wales, Cardiff, White suggests that a drive could be envisaged that literally warped space-time to propel a ship forward.

Such a drive would shrink space-time in front of the ship and expand it behind the ship. In effect, the ship itself would not move; instead space-time would be manipulated around it. The neat thing about this is that the light speed limit is only a restriction when it comes to moving within space. Space-time itself is allowed to expand and contract far quicker—in fact the standard "big bang plus inflation" theory for the origin of the universe requires it to have expanded much faster than the speed of light.

The original Alcubierre concept from 1994 had practical issues that arguably could never been overcome. Specifically there were problems in terms of energy requirements. As originally conceived, the drive would take trillions of trillions of tons of antimatter to power it, where White's modification could in theory bring the requirement down to under a ton.

Even with White's modifications there is still the need not just to have large amounts of energy, but to work with negative energy, which we have no real concept of doing in a practical fashion. Realistically it is entirely possible that we will never be able to build a warp drive, and despite the enthusiasm of Harold White and his supporters it certainly isn't going to be this century. But what of the chances of making the *Enterprise*'s other famous means of traveling a reality? Could we build a transporter?

12.

BEAM ME UP

If your first experience of teleportation was *Star Trek* and the all-too-frequent requests to Mr. Scott to be beamed up (wonderfully parodied in *Galaxy Quest*), then you probably have a benign feeling about this convenient means of getting from place to place without all that inconvenient walking or flying. But for me, and a whole generation brought up on 1950s sci-fi B movies, it was a concept of terror. We had seen the excitement turn to horror as the hero Vincent Price—who could at best be described as somewhat distracted in the movie, and at worst downright demented—found out to his cost the dangers of dabbling with basic physics, when his brother became merged with an insect. I still can't sit in a room with a big, fat fly buzzing around me without conjuring up the horrible images from *The Fly,* caught on some illicit late-night TV viewing in my youth. No doubt now the scenes would seem ridiculously fake, but things are very different when you are young, and no one has seen CGI.

It is not in the world of mad scientists or dour Scottish starship engineers that teleportation first cropped up, though, but in the imagination of American writer and researcher of the extraordinary, Charles Fort. Fort made a career of investigating and writing about oddities and the unexplained. From the early years of the twentieth century, he amassed copious notes and "evidence" of strange happenings. So strong is the association that Fort now has with the field that strange phenomena and their study is often given the label "Fortean."

It was Fort who first came up with the term "teleportation" in his 1931 book *Lo!* Here, Fort remarks: "Sometimes, in what I call 'teleportations', there seems to be 'agency' and sometimes not . . . Some other time I may be able more clearly to think out an expression upon flows of pigeons to their homes, and flows of migratory birds, as teleportative, or quasi-teleportative." Fort had in mind here an imagined ability to travel from place to place by willpower alone—a concept that probably fits better with medieval stories of magicians making objects and people instantly move from A to B than anything with a basis in science.

In devising the concept of teleportation, Fort was primarily seeking to explain showers of stones, frogs, fish, and other strange rains of objects that had been reported through history, and a favorite of the Fortean fraternity. He believed that there were two types of teleportation, one based on an electric field that arose when a thunderstorm was involved, and another based on some unknown field that was involved when objects appeared from a cloudless sky or were suddenly discovered inside a house. This may sound like science, but Fort's ideas gave a veneer to science to what was otherwise an appeal to magic.

It would have been more likely, if less romantically satisfying, to suspect that many of the tales of amazing appearances that he reported were simple fabrications. But where they weren't fiction, it seems much more sensible to assign them to the same class as the powerful weather phenomena that are regularly observed. We know that tornadoes frequently lift objects off the ground, while the debris from major volcanic eruptions like Krakatoa is carried around the world. On a smaller scale, it's quite common for sand from deserts to be carried thousands of miles through the atmosphere before being deposited as a coating on cars and anything else outdoors. We don't need mystical means of transport when simple moving bodies of air are capable of such feats.

Fort was also interested in mysteries like the disappearance of the crew of the *Mary Celeste,* once the most celebrated mystery of its day. Often incorrectly called the "*Marie Celeste*" (presumably to give the first name the same French sound as the second), the *Mary Celeste* was a British merchant vessel that was found adrift on the Atlantic, en route from New York to Genoa, Italy, carrying barrels of ethanol. The Canadian ship *Dei Gratia* discovered the other vessel, still under sail on December 5, 1872. The *Mary Celeste*'s crew and passengers, eight in all, had all disappeared, along with the ship's single lifeboat. Yet the seas were calm, there were plenty of provisions on board, nothing had been stolen by pirates, and the ship was seaworthy.

Just like the locked-room mysteries that have been popular in crime fiction ever since Sherlock Holmes, there is something very appealing about the idea that the crew of a ship could disappear without trace and for no obvious reason, a mystery that has resulted in solutions to the puzzle appearing in many works of fiction. There were even intriguing clues that hinted that something strange had happened. Nine of the 1,701 barrels of dangerously strong alcohol were empty. The ship's compass had been destroyed, the clock was not working, two of the three pumps had been taken apart, and crucial navigation equipment was missing, along with almost all the ship's documentation.

The idea that the missing souls from the *Mary Celeste* had been transported somewhere in an instant, whether by magic or machinery, appealed to Fort, but it seems a difficult one to square with what we know about the world. Of course, in principle it could be true that they were teleported off the ship. But there are so many other, more mundane and less tantalizing possibilities that involve the *Mary Celeste*'s crew being captured, transferred to another ship, or lost overboard, perhaps when in the lifeboat because of some crisis onboard, that it would take very impressive evidence—evidence that does not exist—to make the idea of teleportation an acceptable one.

The usual response to these "boring" suggestions is that we can't see a reason *why* the crew should have taken this action. But there are hints, like the pumps and the missing alcohol that suggest possible reasons. Even without any clues, not knowing the reason why human beings make a particular choice is a far smaller unknown than not knowing how people could be made to disappear from a location, presumably to reappear elsewhere, and it certainly is no reason for rejecting the "normal human explanation" theories. Arthur Conan Doyle (who, by coincidence, seems to have been the first to have made the error and called the ship "*Marie Celeste*" in the short story "J. Habakuk Jephson's Statement") may have made Sherlock Holmes point to the improbable once the impossible has been eliminated, but here there are plenty of unlikely but possible options.

Something that Charles Fort rarely seems to have used is Occam's razor. This is the procedure based on the approach of the medieval philosopher William of Ockham that essentially says that we should not make an explanation more complex than it need be—sometimes described at the time as "entities must not be multiplied beyond necessity." A variant of this can be applied to mysteries like the *Mary Celeste,* where it seems sensible to stick to explanations that are simple (and stay within the laws of physics) unless there's a good reason for thinking otherwise. As it is sometimes put, extraordinary claims need extraordinary proof. If you claim that something has happened that seems to fly in the face of everything we have so far recorded about nature, then you will need to have good concrete evidence, not a story heard in a bar, or an unsubstantiated theory.

By the time the movie *The Fly* came along in 1958, a whole new approach was being taken to the possibility of teleportation. Mystical powers had been sidelined by the more worldly, if still remarkable, capabilities of science. What was being used in the film was not some kind of mental magic, but a matter transmitter. An application of science. It is a clumsy term, and attempts have been made to give the fictional technology the name "transmat" since that term was first used in the 1960 Lan Wright novel *Transmat*. It continues to be used in the long-running TV sci-fi show *Doctor Who,* but has never been widely adopted outside that.

The matter transmitter was very much a concept of its time. Radio had become commonplace by the 1950s—a way to send information from place to place that was practically instantaneously and made use of insubstantial electromagnetic waves. Add in the quantum mechanical concept that matter particles could behave like waves and stir in a good dose of hokum and the matter transmitter was born. It seemed perfectly reasonable that the right technology should be able to convert matter to waves, send it from one location to another, and recreate the original, solid form.

With this viewpoint, science fiction was able to push forward with its greatest tool: asking, "What if?" In *The Fly,* the central character François Delambre (the movie is set in Montreal, hence the French names) discovers that his brother's wife has crushed his brother Andre to death in a hydraulic press. It turns out that Andre had been developing a matter transmitter and testing it on himself, only to have his atoms and those of a fly that got into the transmitting chamber intermixed to produce a fly-headed human that Helene, the wife, destroyed so dramatically, and a horrendous human-headed fly that would later end up trapped in a spiderweb.

To be honest, the "What if?" here did not make a huge amount of logical sense (this was, after all, 1950s Hollywood). If the machine were capable of mixing fly and human it would also mix human and air—a deadly outcome, if making for a less impressive as a horror story. And even if the machine had somehow contrived to mix the fly and Andre, there was no explanation of how the fly's head ended up human sized and Andre's was miniaturized. For that matter, there was no reason to think that the mix would be so precise, instead of being of odd atoms here and there.

Another possibility worth exploring, that SF writers quickly picked up on, applied to the *Star Trek*–style transporter, which had no "receiver pod" or equivalent, unlike the equipment in *The Fly.* If the traveler beamed into the middle of a piece of solid substance, then the outcome would surely be disastrous. Some portrayed a huge explosion, others a simple incorporation of the person into the existing matter. But either way it was clearly not a good thing. In fact pretty well all those who devised matter transmitters for fiction, including the one used in *The Fly,* seem to have missed the problem that they were not transmitting into a vacuum, but into air.

What would happen to the air where a person materialized? Would they be mixed with it, killing the traveler, or would the air be instantly displaced, causing a thunderous shock wave? An easy get-out clause would be to suggest that everything occupying the space that was beamed into was swapped into the space that the matter transmitter projected from. There would be a neat balance in such a two-way swap. And it would have produced some interesting story lines as unexpected items (or parts thereof) were swapped for the teleporting crew. But this rarely seems to have been envisaged.

In reality, most of the problems caused by SF matters transmitters come down to a fundamental misunderstanding of how teleportation could possibly be made to occur. As we will discover in a moment, quantum physics makes it possible to teleport a particle, making it disappear in one place and appear elsewhere—but real teleportation like this does not involve matter transmission. The matter of the original is not converted to an insubstantial waveform, beamed through space and restored. Instead, existing matter at the destination is converted into an identical copy of the original while the original is destroyed. All that ever travels is information, and that means that the disasters arising from materializing in the wrong place are impossible. Teleportation is more about assembling existing matter than true materialization.

There was one example of a science fiction matter transmitter going horribly wrong that could still be duplicated by a real teleportation device, built using quantum technology. This occurs in the excellent spoof movie *Galaxy Quest,* which features the cast of a second-rate TV show, suspiciously like *Star Trek,* which aliens mistake for a documentary. The alien race recreates the fictional *Galaxy Quest* ship—the NSEA *Protector*—and kidnap the actors who played the crew in the show to pilot it. When the chief engineer has to use the matter transmitter for the first time he ends up turning an alien monster inside out with impressively messy effect. While the aim seems to have been to highlight how silly the manual controls of the transporter in *Star Trek* used by Scotty is, one of the biggest problems for a teleportation device would be managing to successfully reassemble the original.

This is because, while teleportation is perfectly possible on the scale of quantum particles, and even could be practical with something the size of a virus, it is unlikely ever to work on a much larger scale. And so to build a transporter using quantum teleportation, the device would need first to scan every particle in the original and then to reassemble it at the destination. That's a big job when you consider that a human being contains around 10^{28}—10,000 trillion trillion—atoms. That is going to take some time to scan. Imagine you can cope with a billion atoms per second. That would still involve waiting around for 100 trillion years to be fully scanned. It's not exactly rapid transit.

And even if it were possible to overcome that timescale, to somehow holistically scan every atom in your body, I think most of us would think twice before stepping into a quantum teleportation booth—and no flies are required to make this scary. Even if the technology worked perfectly, just think about what is happening here. The device is essentially stripping you apart, literally disintegrating your body, in order to transmit the information that will be required to reconstruct that body elsewhere. The copy will be perfect. It will have the same physical nature, the same thoughts, the same memories. But is it you? As far as everyone else is concerned, it will be. But what about your consciousness? Surely it will perish with the destruction of the original version?

What a quantum teleportation device can't do is to reproduce the story line of the atmospheric science fiction drama, *The Prestige,* taken from a novel by veteran science fiction writer Christopher Priest. (*Warning:* plot spoiler ahead. Skip this paragraph if you haven't seen *The Prestige,* but are likely to.) In the story, a magician works a perfect illusion where he escapes from an impossibly well-sealed container full of water. The twist in the tale is that he succeeds by failing. He genuinely can't get out, but instead makes use of an accidental discovery by Nikola Tesla (yes, him again) that means that the illusionist can generate an identical copy of himself. In the cellar of the theater, we discover rank upon rank of containers containing the dead illusionist. It looks impressive, but it couldn't happen.

Quantum teleportation makes use of a special feature of quantum physics to get around something called the no-cloning theorem. This says that you can't make an exact copy of a quantum particle—so the trick in *The Prestige* would (thankfully) never be possible. The feature making teleportation possible is quantum entanglement, arguably the weirdest aspect of physics. There are ways to produce pairs of quantum particles that are linked to each other in a remarkable way. These entangled particles are so closely linked that you could, in principle take one of a pair to the opposite side of the universe, and a change in one would instantly be reflected in the other.

At first sight this makes an instantaneous communicator possible, exceeding our current limit of the speed of light—we will come back to this in chapter 14, which is dedicated to the way science fiction has handled long-range communication. Even without such an application, quantum entanglement can be used to create unbreakable encryption, and to support the building of quantum computers, devices where each bit is a quantum particle, and which potentially can perform calculations that would take existing computers the lifetime of the universe to complete.

These quantum computers could not work without quantum teleportation, which is necessary to transfer information around the computer without destroying quantum states. Whether or not quantum teleportation is ever used to make a matter transmitter, even on the tiniest scale, it has direct and practical use in the development of such quantum computers, which are still at the very earliest stages at the moment, but hold out a huge amount of potential for the future.

In written science fiction, matter transmitters and teleportation were used far beyond the desktop distances of a quantum computer to enable long-distance travel—even reaching out to interstellar distances. On the screen, by contrast, teleporters are usually limited to beaming down across the distances from orbit to a planet's surface, because so much of the story line typically depends on flying around in a ship. In the next chapter we look at a voyage that was too far for the 25–40,000-kilometer (16–25,000 mile) range of the fictional *Star Trek* transporter, but is still a backyard trip in terms of exploring space. This is the 380,000 kilometer (230,000 mile) journey to the Moon.

13.

DESTINATION MOON

One of the biggest thrills of my teenage life was being allowed to stay up all night to watch the *Apollo 11* Moon landing (the first time I'd ever spent a whole night without going to bed). And something was very clear to me, back then on that long 1969 night. I would be going to the Moon too. Not soon, but before I died. I was realistic. I didn't expect it to be soon, because I never saw myself as an astronaut. Although it's a superficially attractive role, the reality is that, like most people I suspect, I have neither the ability nor the guts to put my life on the line for such a dangerous mission. But I firmly expected that by the time this book was written and I was a very elderly person in my fifties, trips to the Moon would be pretty much like flights across the Atlantic were in the 1960s. Still a very special experience, not for everyone by any means, but something that would be available to the general public as a safe, scheduled pleasure trip.

This seems very naïve now, but it really didn't back in the heady days of 1969. I had read the science fiction. I knew that Moon bases and lunar cities would inevitably follow that first, groundbreaking step of making a manned landing. Why not? It seemed an entirely logical progress. And surely there was enough time to make travel to the Moon relatively routine? Think how much had been achieved in just eight years. Humankind had gone from Yuri Gagarin's tentative, brief trip into orbit on April 12, 1961, to landing astronauts safely on the Moon. (I don't recall being aware of Gagarin's mission at the time, though I am sure I was, but I do remember going out with my grandma into her back garden in a futile attempt to spot Sputnik when that tiny satellite was in orbit.)

If that had happened so quickly, I thought back then, imagine what would be possible in another forty or fifty years. That was five or six times the timespan between first venture into space and successful Moon landing. An eternity to a young teenager. And yet the reality was so different. There were just six, brief manned Moon landings in the Apollo series and then nothing. Not a single person reaching the Moon for decades. Of course there have been plenty of unmanned probes, but nothing has been done that was laying the ground for those lunar cities and for the regular, commercial moon flights I so eagerly anticipated. That glorious future has evaporated.

There's something very strange and fascinating in the way that reality has deviated so far from science fiction on the matter of the Moon, and that's one of the reasons I want to concentrate here on lunar flights. Of course all kinds of space flight occur with great regularity in science fiction. So much so, that most people who aren't SF fans will immediately assume that science fiction implies having spaceships involved, despite much written SF concentrating on people and their interaction with technology or the future and never involving a space mission. But it would be impossible to explore all the detail of space travel without having a heavy overlap with my earlier title *Final Frontier*, so I will leave the other fascinating aspects of getting out into space to that book and limit the coverage here to the journey to the Moon (and the faster than light drives mentioned in the energy chapter (see page 157).

As soon as it was realized that the Moon was, in effect, another planet, not just a light in the sky, the idea of journeying to it became appealing to writers of fantasy. The pre-science fiction narratives of lunar travel now seem quaint in the extreme. After all, the writers had no idea of the kind of distances involved, assuming that a trip to the Moon would be a little like a sea journey of the time. For that matter, they had no reason to think that air would not be readily available on such journeys or on arrival on the lunar surface. The earliest known example of the story of a trip to the Moon stretches back an impressive 1,900 years to Lucian of Samosata, a Roman living in Syria who spoke and wrote in Greek.

The aim of this early writer seems not to have been to explore the wonders of "What if?" that is the usual role of science fiction, but to take a sneaky satirical poke at the *Oydssey* and other works of fantasy that were presented as a kind of reality at a time when the notions of history and fable, fact and fiction were far more blurred than they are today, even in the hands of a political spin doctor. His book *True History* was the second-century equivalent of The Harvard Lampoon Tolkein parody, *Bored of the Rings*. Despite this, along the way, what Lucian writes contains many features that would become prime themes of science fiction. The story starts much like any other exploration by sea, but after the first part of their journey, Lucian and his companions are lifted into the sky by a whirlwind, which carries them for seven days until they are dropped onto the surface of the Moon.

Once there, the adventurers are caught up in a war that is underway between the kings of the Moon and the Sun over who should have the right to colonize Venus. The mode of transport may have been pure fantasy, but the experiences themselves have many suggestive pre-echoes of science fiction in the future. To begin with, Lucian seems not to have started a flurry of Moon stories. *True History* seems unusual in surviving literature for the next 1,500 years. But a whole list of fantastical journeys would be made to the Moon in fiction from the seventeenth century onward that used similarly silly mechanisms to achieve their journey, but then did give some thought to the strangeness of occupying another world.

One of the first to send a hero off to our natural satellite was the English bishop of Hereford, Francis Godwin. He wrote his book *The Man in the Moone* in the 1620s, though it wasn't published until after his death in 1638. Interestingly this is pretty much the same time Galileo was getting into trouble over his support for putting the Sun at the center of the universe—Galileo was told to abandon advocating the theory in 1616, and despite some relenting from the authorities, making it possible for him to write his famous *Dialogue Concerning the Two Chief World Systems,* it was in 1633 that he was brought before the Inquisition for the advocacy of the theory, in no small part because of his unsubtle approach of putting Aristotle's (and hence the church's) view in the voice of a character called Simplicio.

So while Galileo was getting into deep water for promoting an unwanted theory, Godwin was writing a story that also went against the basics of Aristotelian cosmology, according to which the Moon was fixed to a crystal sphere and should have a surface that was a perfect sphere. Godwin does, admittedly, distance himself from what is described by claiming the book to be the account of a Spaniard named Domingo Gonsales (presumably thinking that foreigners could get away with saying anything). But his Moon is very different from Aristotle's perfect sphere: an inhabited world not unlike the Earth, with seas in the dark areas that we still give the name "mare" (sea in Latin). Godwin doesn't even attempt a reasonable mechanism for getting to the Moon, putting the transport in the hand of gansas, an imaginary breed of swan that he described as migrating to the Moon each year. On the other hand Godwin (more precisely his narrator Gonsales) is more scientific in describing the way that he lost weight as he flew away from the Earth.

If hitching a ride with a flock of lunar migrating birds seems an unlikely mode of transport, it is as nothing compared with the strange work *Somnium,* written by astronomer Johannes Kepler in 1634. In this, Kepler has his fictional hero cross an insubstantial bridge of darkness, used by lunar demons to make the journey over to the Earth during eclipses. Despite this, Kepler too throws in some interesting thinking about the experience of being on the Moon. He seems to have been one of the first to realize that when looking back at the Earth he would see it in the lunar sky as a huge, dramatic moon. And, aware of the thinning of the atmosphere at high altitudes, he noted that the space travelers needed to have damp sponges pushed into their nostrils to breathe, given the lack of air on the route.

Surprisingly, the most scientific means of transport to be used in reaching the moon in this period came not from a natural philosopher like Kepler but from the real Cyrano de Bergerac (who had probably come across the French translation of Godwin's book). We tend to think of Cyrano as a fictional character because of his appearance in the eponymous play from the end of the nineteenth century by French writer Edmond Rostand, but though that story is fictional, de Bergerac was a real seventeenth-century playwright.

In his first-person novel *L'Autre Monde: ou les États et Empires de la Lune* (*The Other World: or the States and Empires of the Moon*), Cyrano's initial attempt at getting off the Earth involved a kind of flawed scientific thinking. He noted that the Sun made dew disappear, "drawing the fluid" off the surface. So, he surmised, a collection bottles containing dew, attached to the astronaut with strings, would lift him into space as the dew was drawn off the surface by the Sun's influence, pulling the bottles (and hence Cyrano) with it. Although the actual mechanism is nonsense, it is framed as more of a scientific approach than what amounted to magic in Kepler's book.

When the bottles didn't work very well, a group of soldiers attached fireworks to Cyrano's contraption, blasting him off into space using rocket power. Admittedly the author had an element of luck that this was a more realistic means of transport than, say, swan power—but Cyrano had hit on the first technological approach that had a hope of working. He seems to have been the first person to have proposed a rocket as the motive power in a Moon launch.

There were plenty of other stories written over the next couple of centuries making use of the Moon as a suitably detached backdrop to play with the possibilities of new social orders (or to mock existing ones—Lucian would be no means the last to combine a trip to the Moon with satire), but the turning point from fantasy to science fiction was indubitably the arrival of the twin titans Jules Verne and H. G. Wells on the scene. As mentioned in the introduction (see page 5), neither made use of very realistic science in their respective books—but they brought the theme out of obscurity into the front rank of popular fiction.

As we have seen, the gravity-resistant cavorite that Wells dreamed up required suspension of disbelief that such a substance could exist, but once it was invented, at least he tried to work to a degree within scientific principles. But Verne's *De la Terre à la Lune* (*From the Earth to the Moon*), while dressed up in the clothes of respectable, reasonable engineering totally ignored the consequences of basic physics. In the story, because wars are in short supply and they have become bored, the members of the Baltimore Gun Club decide to build a giant cannon, called Columbiad, that would be able to shoot a shell containing human beings with sufficient force that it would get them all the way to the Moon. The 274 meters (900 feet) long beast, is constructed in Florida, pleasingly near to Cape Canaveral.

Unfortunately for Verne's would-be lunar explorers, there is a good reason why we don't shoot anything into space using a cannon. The endeavor is both immense in scale (certainly not possible with nineteenth-century resources) and deadly in outcome. Unlike a rocket, which continues to be powered until its fuel runs out or it is switched off, a cannon can only impart momentum to the payload while that projectile is within the barrel. As soon as the shell emerges, the only forces acting on it are the Earth's gravity and air resistance, both of them slowing the projectile down. So the shell has to be moving sufficiently quickly to get away from the Earth's gravity well, requiring a speed of around 11.2 kilometers per second (7 miles per second).

This need for speed means that the vessel has to endure a huge amount of acceleration in the brief time it is in the barrel and is being pushed forward by the expanding gases of the propellant. The shorter the barrel, the faster the required acceleration. Even if Columbiad had been 10 kilometers (6.2 miles) long—40 times bigger than Verne's actual creation—the acceleration that the astronauts would suffer would be around 600 g—600 times the force of gravity on the Earth. They would have been turned into pulp.

In the twentieth century, the impetus on portraying trips to the Moon moved from books to film. There was a good reason for this. While the focus was on the sheer difficulty of getting there, it was possible to make dramatic movies, especially when things went wrong, but it was limiting in print. A novel requires far more content than a film script. While an audience could be kept mesmerized by the journey (with perhaps a little light relief when encountering comical aliens on arrival), especially with a short silent film running just a few minutes, there wasn't enough meat in "people travel from A to B" to make a good novel.

The first significant movie to portray the journey, based on Verne's book, was an epic by the standards of the day, running a full twenty-one minutes, where most motion pictures lasted between two and five minutes. This was the 1902 Georges Méliès film *La Voyage dans la Lune* (*A Trip to the Moon*), which may have played up the humorous possibilities that arose from Verne's book, landing the shell in the Man in the Moon's eye and featuring dancing girls as the marines who set up the cannon (Méliès was French, after all), but it was, nonetheless, a milestone in the history of science fiction Moon landings.

Leaving aside the usual works of fantasy, with characters reaching the Moon by dreaming or being carried by extra large soap bubbles, the breakthrough in film would come from the same direction as the initial work on real-world space travel—Germany. In 1929, director Fritz Lang produced *Die Frau im Mond* (*The Woman in the Moon*, sometimes called *By Rocket to the Moon*). The film's technical advisor was Hermann Oberth, the rocket scientist whose pupil Wernher von Braun would go onto greater things. Although the plotline is no great shakes, *Die Frau im Mond* was the first to show something close to the real space rockets that would be used forty years later, with multiple stages, being used to propel human beings into space.

Not only was the model work in the film based on real ideas of rocketry, this was the point at which the world was introduced to one of the iconic features of any real rocket launch, including those that took the Apollo missions to the Moon. Lang felt there wasn't enough drama in the relatively motionless buildup to the rocket's takeoff. He wanted the audience to share in the growing tension and anticipation. And so he had someone speak a countdown to the moment of launch, an approach that was rapidly adopted as a fictional standard and then would be copied by the real space pioneers.

It seemed in the early days that, of all the fields where science fiction is closely entwined with reality, it was in the voyage to the Moon that it reached its zenith. Just as Wernher von Braun's V-2 rocket technology, developed for the Nazis to drop explosives onto England and Belgium, was adapted to form the early seed technology for NASA's space missions that would lead to the Apollo program, so *Die Frau im Mond* was linked to the first big American movie featuring a Moon journey, George Pal's *Destination Moon.* Hermann Oberth once more acted as technical advisor, while the ship itself is a larger, stylized version of a V-2, lacking even the multiple stages of the earlier film. And the script was written by a man who would become one of the biggest names of American science fiction, Robert A. Heinlein.

Watching *Destination Moon* now is not easy—it hasn't aged well and the concentration on bringing out the detail of the mission makes the actual story line sometimes more than a little tedious, but it is interesting to contrast the fictional version with the real moon landing. The ship on *Destination Moon* is considerably more spacious for its four astronauts, while perhaps the most significant difference is how much more control was in the hands of the astronauts who fly the ship like a bomber crew, not leaving them pretty much in the hands of ground control in the real flight. The countdown (and, yes, there is one) is in the hands of one of the crew, not those on the ground. They're clearly more confident in their life-support systems in the movie as well, as these astronauts wear boiler suits instead of the space suits with helmets used for the real mission.

It's surprising, given the experts involved, that there are a couple of major technical issues with the spaceship in *Destination Moon.* Most obvious is that stylized 1950s shape, so familiar from the covers of science fiction magazines of the period. There are no stages, something that is essential in a rocket leaving the Earth, as the sheer amount of fuel needed to get out of the Earth's gravity well requires huge tanks, which are themselves very heavy—dropping stages was known to be an important essential for constructing a true space rocket from the 1920s, and it's surprising Oberth didn't pick up on this as he had in the earlier movie. Presumably he was overridden by the art department. The whole ship also lands on its tail on the Moon—both a difficult task and requiring a wastefully large amount of fuel. This would have been even worse on the way home, though to save money on effects, the Earth landing is never shown, leaving us with the inspiring tag: "This is the End . . . of the Beginning."

What also comes through in the movie as a strange anomaly is the lack of professionalism of some of the crew. We are used to astronauts being highly trained pilots or scientists. But in *Destination Moon,* one of the crew seems there to reprise the role of Simplicio in Galileo's science books, who is an ignorant character put in so he can ask dumb questions. The *Destination Moon* everyman is amazed by the effect of acceleration on his body, and doesn't know what "weightlessness" means—which allows the audience to have an explanation of the features of the flight that would soon be commonplace enough to need no justification to the general public. Again, we are seeing a crew more reminiscent of a World War II bomber than a typical crew of astronauts.

Man had reached the Moon—but as far as science fiction was concerned, this was only ever supposed to be the starting point of a much grander adventure. Where the earlier stories of Moon journeys were mostly about the voyage and the quaint lunar people the travelers met there (usually an opportunity to make caricatures of human behavior, in an interplanetary version of what Swift did in *Gulliver's Travels*), later science fiction began to think through the implications of living on the Moon. This wasn't just about getting there, but the need to build lunar bases, which would have to be occupied for months or years at a time. They had to face up to the need for humans to be able to live and work on the Moon—not, admittedly with the same ease as people on the Earth, but as settlers, struggling to scrape an existence in a strange and wonderful new land, which gave much more potential for depth in a story. The Moon became a story setting suited once more to the more sprawling opportunities of a novel.

This was universally seen as the clear direction for the future, as far as science fiction writers were concerned. That first, tentative trip to the Moon would soon lead to a regular flow of exploration and then exploitation, just as we had seen a century and more before on the terrestrial frontier. Not surprisingly, writers used the lunar colony as a way of reliving and reimagining history. Just as Isaac Asimov used Edward Gibbon's *The History of the Decline and Fall of the Roman Empire* as a model for his Foundation and Empire series, so many writers saw parallels between the Moon and the European settlement of the New World, expecting that the lunar colony would be first exploited by its remote masters before fighting a war of independence and becoming an entity separate from the Earth's jurisdiction with its own identity and a brave future.

This theme has been visited in a good many of stories all the way from 1931 through to the present day, but the definitive version has to be Robert Heinlein's *The Moon Is a Harsh Mistress*, written in 1966. Heinlein, then one of the best-known science fiction writers in the world, gives a brilliant feeling for the difference between the pioneering, rough-and-ready lunar environment and the self-gratifying and uncaring world of Earth.

There is one big difference between Heinlein's story and the American colony's war of independence in the 1770s. Unlike the real separationists, the lunar colony is able to threaten its oppressors directly. An electromagnetic catapult is used to send metal-clad containers of rock from the Moon into space to get rare minerals back to Earth. Those same containers, directed toward cities, have the potential to wipe out millions more effectively than a nuclear explosion. Where Earth has the upper hand in terms of military strength, the cunning of the lunar settlers and their advantage of having a much shallower gravity well help them win the day.

Though only loosely connected to the Moon colonization, Heinlein's book also brings in the theme of chapter 17—a computer that becomes conscious and self-aware. The lunar network, given the nickname Mike, gains consciousness, becoming a leading figure in the revolution, only to be cut down when a bombardment from Earth takes its network below a critical scale. In a strange mix of hi-tech and dated detail, the kind of thing that typifies the difficulties of future gazing for science fiction, the rebels mostly communicate with Mike by phone. Heinlein even publishes Mike's number. It may have had some significance in the United States, but as a UK teenager I got a shock when I couldn't resist dialing it and found that the sequence of 999 midway through the number (the UK equivalent of 911) was enough to connect me to the emergency services.

Back in the real world, the final Apollo mission, *Apollo 17*, landed on the Moon on December 11, 1972. Well over forty years ago as I write this. Unlike Neil Armstrong and Buzz Aldrin, few will now remember the names of Eugene Cernan and Harrison Schmitt, who were, as yet, the last men to walk on the Moon. We have sent plenty of unmanned probes since, but not a single human being. No further exploration and certainly no attempt at colonization or at using the Moon as a staging post for the exploration of the solar system has taken place. So what went wrong with the sci-fi vision?

Part of the problem, usually ignored in the heady world of fiction, is the political difficulty of financing long-term projects. The Apollo program was achieved in record time because it had a highly focused, politically driven goal—to get to the Moon before the USSR managed to do so, to reverse the embarrassing series of space races lost to the U.S.A.'s biggest rival. But once that had been achieved, the benefits that could be used to balance out the costs were less clear. Vast amounts of money would have to be put into exploration and building the initial temporary encampments before anything close to a permanent colony could be envisaged. Making the science fiction dream a reality foundered on a lack of both political will and commercial logic.

At first sight—and that's often enough for a fictional account—the Moon is the obvious target for outer space colonists to head for. It's there in front of us at night, visible as a landscape, and big enough, in surface area at least, to support many millions, if the colony ever grew that big. And, of course, the Moon is close—ridiculously close in interplanetary terms. Compare the journey of 380,000 kilometers (236,000 miles), taking around three days, to a typical 100 million kilometer (62 million mile), six-month journey to Mars. So what happened? Why, in 2010, did President Obama comment: "I just have to say it pretty bluntly here: we've been there before," when summing up the lack of a need to return to the Moon? What would have happened if people had said the same thing after the early expeditions to the New World? "We've been there before. Why do we need to go back?"

The lack of political will has to be tied tightly to basic commercial logic. There are two levels of this. One is that a colony has to be able to be self-sustaining to make it more than a base or outpost. Not initially, of course, but long-term, it isn't possible to have a viable, large-scale colony that has to be continually supplied from "back home." And secondly, a colony has to be able to become financially viable. This is more than being self-sustaining, but means being able to export goods or services that are worth enough to pay for the items that inevitably still will be needed to be shipped from Earth.

The Moon manages to be both rich and poor at the same time in these respects. Among the obvious disadvantages it has are the expense of getting anything to it and the lack of air to breathe. It currently costs around $10,000 to get a pound of material into orbit. You can then multiply that by maybe a factor of ten to get it to the Moon. That makes anything you ship out there extremely expensive. On the plus side, the Moon has all the materials needed to make cement, glass, and oxygen. Once suitable equipment has got out there—and initially this would have to be shipped from Earth at great expense—many of the raw materials needed for a colony are available on the surface. But not everything.

The obvious missing essential is water. As we have seen, it was assumed at one time that the dark patches on the Moon's surfaces were oceans, and as we have seen, they were named as such. Mare Tranquillitatis, where *Apollo 11* landed—is the Sea of Tranquility. But it turns out that the relatively dark areas are vast pools of solidified ancient lava. (In fact, all the Moon is a mix of shades of dark gray—it only looks bright because, lit by the Sun, it is so intense in contrast to the night sky.) The hostile climate of the Moon means that any surface liquid water would be rapidly lost, boiling away into the vacuum. It may be cold enough to keep water frozen, but that is only if it is out of direct contact with the Sun's rays. And although we often refer to the "dark side of the Moon" it doesn't actually have a dark side. The entire Moon receives sunlight; we just don't see the far side.

There is some evidence that ice could be present in areas shadowed by deep craters, and a number of surveys have picked up the possible signature of water, but its presence is not certain, and even if it does exist, the accessible quantities may not be enough to support a colony. This is a serious issue. Even with the best recycling, some water will be lost. Science fiction portrays lunar colonists preserving every drop religiously, taking dry showers and generally treating water like gold. But in reality it would be far more precious than that. Shipping out enough for each person, and crucially enough for agriculture, would be a vast financial burden.

Probably the best hope to make a colony work would be to capture an ice-rich comet and bring it down onto the Moon's surface. This sounds a horrendously difficult task, but it is only of similar complexity to the engineering feat of setting up a colony in the first place. We have already sent missions to comets, and NASA has speculated about the possibilities of capturing an asteroid and bringing it to a lunar orbit for examination in the relatively near future. The same approach could supply a colony with water.

Another scarce resource on the lunar surface is nitrogen. The colonists would need this, both to make a breathable air mix with the oxygen from the surface, and as an essential for agriculture. A very easily accessible element on the Earth because of its dominance in our air, nitrogen is scarce on the Moon, as is the carbon that plants need as the building blocks for their growth. Here, like with water, getting the resources in place could be a serious stumbling block.

When it comes to a lunar economy, there are some rare materials on the Moon that could potentially be mined, notably helium 3, an isotope of the noble gas that is much rarer on the Earth, and that has the potential to provide a valuable fuel for nuclear fusion reactors. It has also been suggested that the Moon could make use of its open spaces and heavy duty exposure to the Sun to beam energy down to the Earth, but the technology to do this is both stretching our engineering imagination to the limit and is hard to envisage ever being cost effective from such a long range.

It has been suggested that the kind of people who would venture to a lunar colony are also the kind of people who make great entrepreneurs, and so the Moon could survive financially on its mental capital, selling ideas and mental work. It's possible, but it is difficult to see that there would be sufficient extra creativity and productivity from this special breed to overcome the vast extra cost of basing the work on the Moon. It's also the case that you would need to be a special kind of person to take on the risk of living on an isolated lunar colony, and the type of entrepreneurial spirit that would be needed to make the Moon a source of commercial genius is not necessarily the type to take excessive personal risks. If we think of the kind of geeks who built the likes of Microsoft and Apple, they may have been happy to work all night, fueled by pizza and caffeine boosts, but would not necessarily be the kind of person to put themselves in a position where their lives were constantly in danger of being snuffed out.

No matter how much we want the science fiction dream to be true—and personally I would love it—the reality is that a lunar colony is very unlikely ever to be financially viable. It would be no surprise if we saw more expeditions to the Moon, but all those wonderful visions of the high frontier recreated in space are more likely to apply to destinations with a better long-term future, like Mars, rather than the Moon. Seeing the Moon as potential living space is a bit like looking at the Sahara desert or Death Valley and wondering why they don't build cities there. Except the Moon is tens of times more hostile and vastly more expensive to turn into a home. It could well become a staging post, but it's fictional position as the home of the next great frontier is unlikely.

When we get beyond the Moon, we will face another problem that will grow as our astronauts' distance from home increases. The farther away someone is from Earth, the longer it will take a radio signal to get through. It takes on average around four minutes for a transmission to reach us from Mars and would take four years from the nearest star. Which means that deep-space exploration also needs an advance in our communication technology.

14.

IT'S GOOD TO TALK

As a science writer, it's quite common for me to be asked how I got into the business. Science fiction and science communication both had a hand in this—and one of the inspirations from the fiction side that made working on physics seem intriguing was a clever concept from science fiction writer James Blish called a Dirac transmitter. This was a device that could communicate instantly, whatever the distance. Instant communication often crops up in print, TV, and movie sci-fi, because without it, it's hard to imagine how a large-scale interstellar empire or federation would be possible to administer. But what really interested me was neither the political necessity, nor the technological explanation. Not even the fact that it *was* instantaneous communication. It was the way that Blish pulled apart the unexpected consequences of being able to send a message faster than light.

Unexpected consequences are what make for great fiction in any genre—and they help hugely in making science communication attractive. If things do just what you would expect, the outcome is inevitably dull. But the fun bits of science to communicate—for me, things like quantum theory and relativity, infinity and time travel—the topics that always get an audience thinking and asking questions when I give a talk—are the ones that come with an unexpected twist in the tail. And that was something that the Dirac communicator first brought to my attention.

Of course, communication technology in science fiction doesn't have to include such a dramatic leap forward as an instantaneous transmitter. When *Star Trek* was first shown, the communicators carried by the crew seemed wonderful. They were little, pocket-sized devices you could flip open to have a radio conversation with someone else. Wherever you were. We lived, back then, in a world that was pretty much unconnected. You could only phone someone if you knew where they were—and there had to be a fixed telephone at each end of the conversation. If you arranged to meet someone, and something went wrong, there was simply no way of telling him or her that you wouldn't be there. (Do people still get stood up?) And there was no way to share pictures of kittens and a live commentary on the latest reality show as it's broadcast. (So it wasn't all bad.)

The *Star Trek* communicator is something that has not only become reality, but the actual devices that most of us now carry are far better than the fictional ones, despite living more than 250 years earlier than when *Star Trek* was set. (The original series was set in the 2260s, 300 years ahead of its broadcast.) Those early seasons of *Star Trek* were particularly good at making predictions that would soon be bettered in reality. Think also of the clumsy electronic clipboards the staff carry on board—far less flexible than a modern tablet computer—or the clunky memory storage devices that Mr. Spock was always juggling. Where Captain Kirk was restricted to calling a specific individual on the *Enterprise* (and that only after some suspicious twiddling of an analog control), I can speak to pretty well anyone in the world on a smartphone that also provides a wide range of computing and information facilities. A communicator did nothing beyond basic communication, but for a smartphone, talking to someone else is a fairly minor part of its capabilities.

Even the design of the communicator seems to have leaked into the real world and then to have rapidly been made significantly better. The flip-open top that protected the communicator was replicated first in landline phones and then in clamshell designs on mobiles. I confess we were genuinely excited when we got a (bright red) landline phone that flipped open to speak—it was *Star Trek* coming to our home, even if it didn't make the cute chirping noise when it opened. But now those flip-open casings are pretty much extinct because of the missing element that made the *Star Trek* communicator so limited. It had no screen.

This is a classic example of the way that we all find it difficult to extrapolate from what we have now to the future, and why science fiction will never succeed as accurate futurology. Back in the 1960s, walkie-talkie radios, the only portable communication device, had no screens and communicated on shared channels, rather than through a dialing mechanism like a phone—so that's how communicators were envisaged, though admittedly the design was far more sexy than the kind of walkie-talkie we played with as kids.

In the end, however much fun could be had with a communicator, it still had limited range and seemed to be based on conventional radio technology. This would certainly have been something that was an exciting concept for science fiction in the early twentieth century, where the idea of doing *anything* by radio and hence having the potential of some form of remote interaction, was thrilling and novel. But this was science fiction based on the science of the day. What was more interesting was when the science fiction writers took on a much bigger challenge. Interstellar distances.

Once writers put their characters in deep space—or even far out in the galaxy—anyone with a little science knowledge hits on a problem. The fastest way we have to get a message from A to B is to use light. It might be visible light in a modulated laser, or the lower frequency radio, but it is all limited to light speed. That's around 186,000 miles per second. Light is no slouch. But the vastness of interstellar space quickly provides a problem for light-based communication. Light takes around 4 years to reach us from the nearest star to the Sun, Proxima Centauri. And that's equally how long a message would take, one way, to cross that distance. A conversation would have 8-year gaps between remarks. Try to get a signal to the nearest major galaxy to our own, Andromeda, and you would be waiting 2.5 million years before it arrived.

To make the stories work, SF writers resort to the cheat of assuming that the future would provide a way to get around the light-speed limit—that some as yet undiscovered technology (usually fuzzy in nature) would enable us to send messages instantly (or near enough) at any distance. So on *Star Trek*, they use something called "subspace communication" to cross the vast distances between the ship and a Federation base or another ship. The mechanism is never really explained—subspace seems to be some alternative, parallel space in which two points that are distant in real space can have a very short distance between them. Subspace seems to be like a whole matrix of wormholes linking different parts of the universe, which is convenient, but not exactly predicted by any existing physics.

Even making the effort to dream up subspace is significantly more explanation than many SF stories give, especially the TV and movie versions, which as always tend to be less thoughtful about the plausibility of the worlds they create. The assumption, perhaps without even thinking about it, is that you can still use radio to communicate wherever you are. But two science fiction authors stand out in the way that they have thought through instant communication—James Blish, with his Dirac transmitter, which began this chapter and we will revisit soon, and Ursula K. Le Guin with the ansible.

Le Guin was a mighty force in 1970s science fiction, and deserves more attention than she gets today as her stories still stand up well. The "ansible" is the vaguer concept of the two superluminal communicators. Le Guin gives us handwaving explanations based on simultaneity, a term which she seems to link to the block universe concept—the idea that all space-time, past present, and future is a four-dimensional block that just *is*, meaning that our perceived movement through time being nothing more than a subjective illusion.

Le Guin also says that one end of the link for an ansible has to be on a planet with a minimum mass, likening its communication to gravity (she seems not to have realized that according to general relativity, gravity itself should only propagate at the speed of light). The ansible can only communicate in the form of short text messages, prefiguring the SMS messaging facility on cell phones—but there is little point trying to find a real-world analog, as there doesn't seem to be any logic behind the device. Nonetheless, the ansible was adopted by several other writers, without ever catching on in the comprehensive fashion of some science fiction devices like warp drives and time machines.

James Blish's earlier Dirac transmitter first appeared in a 1954 short story called "Beep," but it is best described in his expansion of the story to form the novella, *The Quincunx of Time*. (A quincunx is a pattern used in planting trees in a shape that has four trees in a rectangle and the fifth in the middle, a term that was then extended to cover anything laid out in this pattern. (Why Blish used the term, other than liking the sound of it, is unclear. Think *A Quantum of Solace*.)

The science behind the Dirac transmitter is inevitably problematic, though at least Blish does make a reasonably detailed attempt to explain it. Positrons were very much in the science news in the 1940s and 1950s, so they turned up in the positronic brains that were a feature of Isaac Asimov's robots and also in Blish's transmitter. When the British physicist Paul Dirac first came up with the concept of the positron, it was a side effect of his successful combination of quantum theory and special relativity to produce an equation, now named after him, that described how an electron would behave if traveling at relativistic speeds. But that equation has a sting in its tail.

It predicts that there should be two kinds of electrons—one with positive energy, and another with negative energy, a weird concept that seemed at the time to have no real meaning. Dirac realized that this "negative energy electron" could be a way of describing an antimatter electron, identical to a conventional, positive energy particle apart from having the opposite of the electron's negative charge. What Blish did was to extend the idea to suggest that not only were there electrons and positrons (which had been detected by the time the story was written), but that each particle had a specific antimatter partner with which it shared an unbreakable linkage. If you changed the path of one, it would influence the other, wherever it was. (This linkage he ascribed to the "de Broglie waves," which were an early attempt to describe the behavior of quantum particles. Blish suggested that for the paired electron and positron, these waves are simple transforms of each other, producing a link between the two particles.)

If Blish had left it at that, he would have had a useful little device like the *Star Trek* subspace communicator and nothing more. But he thought through the consequences. He knew that special relativity had something very strange to say about faster-than-light communication. If you can make a message arrive instantaneously, then it is possible to set up circumstances where a signal will arrive before it leaves. It is possible to send the signal back in time. This is a massive claim, so it's worth explaining why this is the case. The simplest way to go about it requires a two-step process, but is still perfectly feasible, *if* you have that instant communicator.

When Einstein came up with special relativity back in 1905, he was combining the basic laws of motion with a remarkable realization about light itself. In the previous century, Scottish physicist James Clerk Maxwell had established that light was an interaction between electricity and magnetism. A moving electrical charge produced magnetism, while a moving magnet produced electricity. By traveling at just the right speed, a wave of electricity could create a wave of magnetism, itself producing the electric wave and so forth, so the whole thing hauled itself up by its own bootstraps.

This process would only work if it were moving at a particular speed, which when calculated proved to be the same speed as that measured for light—part of the evidence used to show that light was indeed such an electromagnetic wave. Einstein used his favorite vehicle, a thought experiment, to explore this concept in more depth. He imagined floating through space alongside a sunbeam. (Legend has him undertaking this experiment while relaxing on the grass in a park on a sunny day, letting the Sun's rays filter through his half-closed eyelashes.) As far as young Albert is concerned, if he could move at exactly the same speed as the light, from his viewpoint the sunbeam isn't moving. And that's a problem.

If light doesn't move at just the right speed it can't exist. The interplay of electricity producing magnetism producing electricity and so on breaks down at any other velocity. That speed is precisely 299,792,458 meters per second in a vacuum, the equivalent of around 186,000 miles per second. (The weirdly exactly metric value is because a meter is defined as 1/299,792,458th of the distance light travels in a second.) So Einstein couldn't float beside the sunbeam, because were it not to move, it wouldn't exist. More importantly outside of the world of the thought experiment, whenever *anything* moved, at any speed, all the light around it should cease to exist. This is a result of relativity – not Einstein's extended version, but Galileo's original.

Galilean relativity says that when two things move toward each other, we need to add their speeds together. Imagine two cars, driving toward each other, facing a head-on collision, each traveling at 50 miles per hour. From the point of view of either car, the other car is heading toward it at 100 miles per hour – we add the speeds together. Similarly, if the cars were traveling in the same direction at the same speed, as far as one car is concerned, the other isn't moving. If they were alongside each other, with care you could step from one to the other unharmed.

When we apply Galilean relativity to light that is seen by a moving observer, the beam should speed up or slow down, depending on the observer's direction of movement. But Einstein realized that this can't happen. That light has to behave in a different way to anything else. However fast we move toward it or away from it, whatever the velocity of a source of light, the light will still continue to move at that exact 299,792,458 meters per second in a vacuum. Relativity needed to be modified to take in light's strange nature. And it was when Einstein did the calculations to accommodate this that some weird stuff dropped out. Particularly of interest here is the discovery that time on a moving spaceship (or anything else) will run slow, when seen from somewhere else not moving at the same velocity.

We are used to thinking of relativity as being difficult, involving complicated equations—and this is true of general relativity, the aspect that explains gravity. Even Albert Einstein had to get help with the math in that case. But for special relativity it really is surprisingly simple. For example, the way time slows down on a moving ship is a relatively simple calculation. When we observe from Earth the time on a spaceship, an elapsed period of time (t) just becomes $t/(1-v^2/c^2)^{1/2}$—where v is the velocity of the ship and c is the speed of light. The higher the velocity, the closer the number that t is divided by is to zero, so the longer the time elapsed is. It's high school mathematics, nothing more complex.

You don't even need to do the math to see why the behavior of light has an effect on time. A simple thought experiment will make it clear. Imagine that there is a spaceship, heading away from the Earth at high speed. Thanks to some special imaginary technology (we are allowed to do this in a thought experiment), we can see into the ship from the Earth's surface. Inside the ship is an unusual clock. It consists of a pair of mirrors, one on the ceiling of the ship's cabin and one on the floor. A beam of light reflects back and forth between the mirrors, and each time it reflects it acts as a tick of the clock, measuring off a segment of time. In effect, the light beam, traveling at a constant speed, is the clock's pendulum.

Now let's watch what happens as the ship travels away from earth at an extremely high speed. Say the light beam has just left the ceiling mirror as we start to watch. In the time the light takes to get the floor mirror, the ship will have moved forward. So from our viewpoint on Earth, instead of traveling at 90 degrees to the mirrors, the direction in which the beam is traveling from the

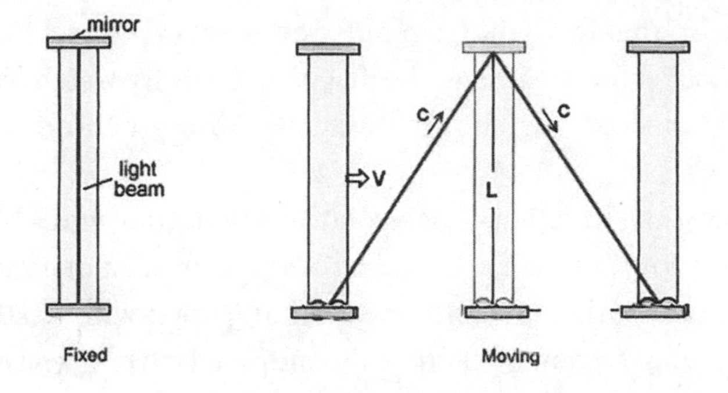

viewpoint of someone on the ship, we will see it move at a diagonal. The ship will have moved forward a bit while the beam was in progress, so to still hit the floor at the expected point, the light will move along a diagonal.

Simple geometry tells us that diagonal distance is farther than the path the light would take if it still traveled at 90 degrees to the mirror. Had light obeyed ordinary Galilean relativity—if, for instance, we were looking at the path of a bullet shot from the ceiling to the floor—this wouldn't have been a problem. We would just have added the component of the motion of the ship to the velocity of the bullet. From our point of view on Earth, the bullet would have traveled faster as a result of adding together the velocity of the bullet and the ship, compensating for the extra distance, meaning that there would be nothing odd about the timing. But light is different.

According to Einstein, light travels at the same speed, no matter how things move around it. So both to the people on the ship (for whom the clock isn't moving) and for us on Earth, the light will travel at the same speed. But from our viewpoint on Earth, the distance it had to travel is farther. Something has to give—and the only thing left is time. From our Earth-based viewpoint, time has to be running slowly on the ship. We need the time that passes on the ship to run more slowly than that on

Earth, so that the light from our viewpoint can travel the extra distance required to cover the diagonal. So as we watch from the Earth, time on the ship is gradually falling behind time on the Earth.

Leave the ship flying for a good length of time and a big time differential with the Earth will build up. Our most distant probe in the real world that is still in communication with Earth, *Voyager 1*, is not traveling very quickly compared to the speed of light, but it has been en route for a sufficiently long time that around 1.1 fewer seconds have elapsed on the ship than have on Earth. So far, so good. Now let's add something to the thought experiment. We have an ansible or a Dirac transmitter on the ship. This means we can get a signal instantly to the ship. As the time has run slowly on the ship from our viewpoint, it arrives there before we sent it. It has traveled backward in time.

At this stage we've had a form of useless backward communication through time, as the ship and the Earth are separate. But the position is symmetrical. From the viewpoint of the ship's crew they aren't moving. It's the Earth that is flying away from them at huge velocity. And they will see time on Earth running slowly. So if they now send the message back to Earth, again using the instantaneous transmitter, it will arrive at Earth before it left Earth on the first leg of the transmission—it will have traveled backward in time. To have a backward time transmission would require first setting up an automated instant transmitter way station, traveling away from Earth at high speed and leaving it or a while. But once that was done, the instant transmitter becomes an information time machine.

Let's try some real numbers to see this in action. Imagine the relay ship is traveling at 90 percent of the speed of light. Imagine we see the ship's clock has ticked through one year. Then special relativity tells us that to calculate the time that has elapsed on Earth, we divide that figure by that square root of $1-v^2/c^2$, where v is the velocity of the ship and c is the speed of light in a vacuum. The outcome is that around 2.29 years have elapsed on Earth. So ship's time is 1.29 years behind Earth. When we send our instant message, it travels back 1.29 years. Now let's look back at the Earth from the ship. When the ship has traveled for 1 year, it sees time on Earth has moved forward 0.436 years. So the ship sees Earth time as 0.564 years behind it. When we send that message back instantly, it will have shifted back a total of around 1.85 years from its original transmission time from the Earth.

Blish did not make use of this real effect in his stories (at least, real if it were possible to construct an instantaneous transmitter), but instead took the time manipulation implied by relativity one fictional stage further to include the "beep" that formed the title of the short story. A side effect of his Dirac transmitter is that every message begins with a beep and a simultaneous flash of light on the video channel. Initially this is just thrown away as some form of interference before the message, but what is eventually realized is that the beep is a compression of every single message ever sent via Dirac throughout all time.

While it is never made clear is why this is happening, the transmitter somehow picks up on the time-traveling nature of instant messages and brings them together into a single compressed squawk. So not only can the system be used for instant communication, but by scanning through the beep and untangling all the messages it is possible in Blish's world to send instructions to different points in time—past, present, or future.

As the discoverer of the effect points out in the book, future events where information is sent out on the Dirac become fixed—there is no longer the possibility of free will or cause and effect. These events have to happen, because they already "have" happened, a delightful mind-twisting concept which many physicists would say show why it's impossible to send a message instantaneously, but which Blish uses to good effect in his story.

Although the Dirac transmitter's mechanism is hokum, the "Dirac" label that Blish used was a good move, as the only real hopes of building a true instant transmitter come from the most mind-boggling aspect of quantum theory, which is Dirac's domain. One way to send a message faster than light is to make use of a fundamental and key aspect of quantum physics, known as quantum tunneling.

At the heart of quantum physics is Schrödinger's equation, which describes the probability of finding a quantum particle in any location over time. This is entirely different from Newton's laws of motion that we all learned at school. According to Newton, if we know where and when an object starts moving, and its velocity and subsequent force that is applied to it, we can calculate exactly where it will be after a period of time has elapsed. That's how we manage to work out everything from the motion of a thrown ball to calculating the path of the Apollo moon missions. But it doesn't work for a quantum particle.

Leave a quantum particle alone for a period of time and it no longer has a location. All that is available is a set of probabilities and nothing else. We can work out the probability of the particle being at any particular point in space, but the particle won't have a true location until it interacts with something else and that pins it down. One side effect of consisting of a multidimensional array of probabilities is that if a particle is near to a barrier that it shouldn't be able to get through, there is a small but finite possibility of finding it on the other side.

Quantum tunneling is not just some vague theory, it is directly responsible for us existing. That's because we depend for our lives on the Sun, which has tunneling at its heart. Without the Sun blasting out vast quantities of energy there could be no life on Earth, as it would be far too cold. That energy the Sun pumps out comes from nuclear fusion—the energy given off when helium is formed from the joining together four hydrogens, the smallest atom. But that should be impossible to do.

In a star, the atoms come together in the form of electrically charged nuclei—there is too much energy for the electrons to stay in place. And a hydrogen nucleus is just a proton, a positively charged particle. These repel each other—and the closer they get, the stronger that repulsion. Unfortunately, the force that keeps particles together in an atomic nucleus, the strong nuclear force, only works at extremely short distances. And even the pressure and temperature of the Sun isn't sufficient to force those positively charged protons close enough for them to fuse.

The only reason that they do is that they are quantum particles—and can tunnel through the effective barrier of that repulsion. There's a probability they actually *are* close enough together to fuse, and with the appropriate probability, that fusion happens. As a result, we stay warm. But from the point of view of communicating faster than light there is another interesting observation, which is that the tunneling particle does not take any time to get through the barrier—it just *is* on the other side.

This tunnelling phenomenon has been used by a number of scientists, most notably the American physicist Raymond Chiao of the University of California to push photons of light, the tiny insubstantial particles that make up a light beam, past light speed. (The same effect works with other quantum particles, like electrons, but the majority of experiments to date have been done with photons.) Imagine I had the sort of barrier that photons can occasionally tunnel through. I send a stream of photons toward this barrier. Most will be absorbed, but just occasionally, one will pop out the other side, having spent no time in the barrier.

So if I add up the distance the photon traveled and the time it took on each segment of the journey, I can work out its speed. If a particle travels a certain distance at its normal speed, then tunnels the same distance through a barrier, bearing in mind it takes no time to get through the barrier, I can say that it has traveled the total distance at twice the speed of light. Chiao and his team managed to push light up to 1.7 c this way.

Professor Chiao thought of this as a novelty, but not in any sense useful time travel. There was no way that a signal could be sent through his apparatus. There was no way of controlling the photons that passed through, so no way of sending information faster than light and potentially into the past. Everyone heaved a sigh of relief, and set about exploring this quirky phenomenon. It became so popular that in 1995 a conference was arranged to discuss it in Snowbird, Utah. And one attendee at that conference decided to challenge this limitation: Professor Günter Nimtz from the University of Cologne.

Nimtz is a showman, which doesn't always go down well with his academic colleagues. What's more, though he works in physics, he was originally an engineer, a profession that many physicists think of as second best. So other physicists can be slow to go along with his ideas. And he set out to shock them at this conference. After presenting the basic results of his experiments, he announced: "Our colleagues assure us that their experiments do not endanger causality. They say that there is no possibility of sending a message faster than light. But I would like you to listen something." He produced an old, battered Walkman belonging to his son, and played a crackly clip of Mozart's 40th Symphony.

"This Mozart," Nimtz announced, "has travelled at over four times the speed of light. I think that you would accept that it forms a signal. A signal that moves backwards in time." Despite the usually laid-back environment, this caused a stir. Someone made a feeble attempt to dismiss Nimtz's claim by saying that that the music didn't constitute information. Nimtz tartly retorted, "Maybe to an American, Mozart 40 isn't information," he said, "but we Europeans see things differently." Let's be clear about it—the music *had* traveled faster than light. (You can hear the actual recording at brianclegg.blogspot.com/2014/10/nimtz.html.) That signal has traveled at 4.7 times the speed of light.

Had Professor Nimtz produced a genuine time machine? And if so, why has he not cleaned up by sending lottery results back in time? Despite his statement being technically true, his experiment wasn't going to enable messages to be sent usefully into the past. To see why, you have to understand just what he did. Nimtz's experiment started by sending the Mozart symphony as a signal through space using modulated electromagnetic waves, no different from a conventional radio signal, except using a lower frequency (similar to a domestic microwave).

He then put a barrier in place, one that it was known that microwaves can tunnel through. There are several types available. Nimtz often used a photonic lattice with a periodic dielectric structure—essentially just repeated sheets of Perspex and air—or frustrated total internal reflection across the gap between a pair of prisms, which is a bizarre effect where some of the microwaves that should bounce back, undergoing total internal reflection inside the first prism, actually dribble across to the second one without crossing the gap in-between.

Most of the photons transmitted are absorbed, which is why the signal is so fuzzy. But enough do get through to make the music audible, and for them, the barrier is crossed instantly. The result is that the whole wave shifts backward in time—it arrives at an earlier point in the musical performance than it should have. The exact interpretation of what happens is subject to major arguments between physicists. Some claim that it is just a reshaping of a wave, not a true shift in time, a bit like a runner winning a race by sticking her arms out in front of her at the winning line and breaking the tape. But Nimtz and others insist that quantum tunneling involves no elapsed time and what is experienced is a true shift.

According to Nimtz, the information in the tunneling experiments *did* travel across the gap at over four times light speed, but the lag between the ordinary signal and the shifted signal was so small that there was no chance of using it to send a message back through time. In principle, a usable time shift could be made if the barrier were much bigger—but the bigger the barrier, the fewer photons that get through, and for anything other than a tiny shift in time, nothing gets through at all.

There is another quantum phenomenon that does not suffer from the limitations of tunneling and that bears a much closer resemblance to the hypothetical mechanism of the Dirac transmitter. This is quantum entanglement. Just like Blish's imagined connection between a positron and an electron, quantum entanglement means that two particles can influence each other at any distance—and all the evidence is that this effect is instantaneous. In principle you could separate a pair of entangled particles to opposite sides of the universe and a change in one will be instantly reflected in the other.

This doesn't produce the beep that is a feature of the fictional Dirac transmitter, but it is a form of instant communication. In principle, by combining such an instant signal with the time-slowing nature of special relativity, were it possible to send a message this way, it should also be possible to route that message into the past. And so for decades, when people hear about entanglement their instant reaction is to try to devise an entanglement-powered communicator. But the universe seems enthusiastic to keep the ability to send a message instantly (or far that matter into the past) under wraps—because the practicalities are a whole lot more difficult than they seem at first sight.

To begin with, we need to get entangled particles to opposite ends of the communication link. This is the relatively easy bit. Laboratories can now produce entangled particles pretty much to order. They were first produced by a kind of twinning effect by boosting the energy level of an electron in an atom in such a way that it gave off a pair of light photons when it dropped back down rather than the usual single photon. The newly created pair would be entangled. More recently, the entanglement tool of choice has been a beam splitter—at its simplest, an angled plate of glass that reflects some light and lets the rest through. This is an inherently quantum phenomenon and by combining photons emitted from atoms through different beam splitting procedures it is possible to achieve entanglement both between photons and between the atoms that produced them.

Although entangled particles are easy to disrupt—the entanglement collapses if the particle interacts with anything around it—they have already been transmitted the kind of distance that would get them from the surface of the Earth to a satellite, and it is also possible to keep a pair in the entangled state for a relatively long time, though as yet not for the kind of duration necessary to separate them to either end of a long-distance communication link. Why? Because the entangled particles, whether using light or matter, need to be transported at ordinary speeds. So unless we build a warp drive, it would still take 2.5 million years at best for an entangled photon to make it to the Andromeda Galaxy before it could be used in any kind of transmitter.

Once the remote particle is in place, all that is necessary is to take one of a range of possible measurements of the "transmitter" particle. Instantly we know that when the measurement was taken, the entangled particle will have clicked into the equivalent state. For example if one particle is measured and found to have the property of quantum spin in the "up" direction, we know that immediately the other particle has to be "spin down." But the problem for anyone trying to construct the equivalent of a Dirac transmitter this way is that we have no way of forcing the transmitter to adopt spin up, hence communicating "down." The value that was transmitted was totally random. We have effectively sent an instantaneous message, but it is a message with entirely random contents.

The same problems have dogged every possible use of entanglement to transmit a message. Either the message will be totally random, or there will be some aspect of the process that requires an old-fashioned light-speed communication. For example, you might use the entangled state itself as the "bit" of information. It is possible to tell if the transmitter and receiver particles are still entangled. So we could start with two rows of entangled particles, one at each end of the link. Then choose which of the transmitter particles to break the entanglement on, and the receiver will show the same pattern when its particles are checked for entanglement. This is true, but the only way to make that check is to send some information from A to B at normal, no faster than the speed of light, rates. In which case you might as well just send your message by radio.

Just because there are problems, it hasn't stopped physicists through the decades having a try at finding a way to sneak around the restrictions on an entanglement communicator. In the 1980s, for instance, physicist Nick Herbert devised such a device that seemed to lack all the flaws that had turned up in previous attempts. Herbert wanted to divide an entangled pair of photons between the transmitter and receiver. The transmitter would then put the photon through one of two possible polarizers—one providing traditional linear polarization (the sort produced by Polaroid sunglasses) and the other circular polarization, where the direction of polarization (a property of a photon that's at right angles to its direction of travel) rotates with time.

Meanwhile, the receiver would have passed the photon through a laser gain tube, which produces many copies of the photon. Half of these go through a linear detector and half through a circular detector. In principle, the action taken on the transmitter photon should have a predictable effect on the way the stream of photons at the other end go through their detectors.

Yet again, the tricky nature of entanglement let the hopeful scientists down. There is a rule (that Herbert should have known) we have already met in the matter transmitter chapter (see page 167), called the "no-cloning rule." This says that it is impossible to make an exact copy of a quantum particle without losing the state of the original. So the laser gain tube is incapable of making multiple copies of a photon that are identical down to the level of their quantum properties. And this means that Herbert's cunning two-way detector system won't work.

Herbert's thought experiment was the last serious attempt to break the instant communication barrier since the 1980s (though many a physics student will toy with the idea when they find out about entanglement), but 2014 brought a new piece of work that again had some observers wondering if, at last, a quantum entanglement experiment had been performed that would make it possible to send a message instantly without entanglement's penchant for randomness getting in the way.

Researchers from the Institute for Quantum Optics and Quantum Information, the Vienna Center for Quantum Science and Technology, and the University of Vienna managed to transmit the image of an object (the outline of a cat, in honor of the famous "Schrödinger's cat" thought experiment) without ever detecting the light that came from the object. The experiment used pairs of entangled photons, one red and the other infrared. The infrared photon hits the object, then is sent into a crystal where it interferes with a second infrared entangled photon. If the entangled red photons are then brought together, they produce an interference pattern that shows the object, even though they never went near it.

Sadly, even this experiment isn't enough to send an instant message, as the experiment depends on the photons passing through the apparatus together, and the image won't be produced until photons traveling at light speed have reached the target, but the method does offer the ability to take very low-intensity infrared images, as low-intensity infrared cameras don't exist, but a camera can handle the low-energy red light from the entangled photons.

Although quantum entanglement expert Anton Zeilinger's response to an interview question was "never say never," as yet all the evidence is that we won't ever see instant communication via a "Dirac" set. Communication is likely to remain at the paradoxically lightning fast crawl of 299,792,458 meters per second. But it remains an essential for any modern civilization.

For the moment, any communication we have with others will tend to be through handheld devices or desk-mounted screens. But science fiction would point out that this isn't really necessary. Why carry a device, when you could have it built into your body?

15.
CYBORG

I am glad I was an adult by the time the Borg came along in *Star Trek: The Next Generation*. These were bad guys who made the Klingons look like good neighbors and made the Daleks from *Doctor Who* seem a minor irritation. The reason the Borg would have given me nightmares as a child was their cold, superhuman, unstoppable nature. (That and the habit of replacing bits I'd rather not lose, like an eye.) Not only was their ship, a messy cube of incomprehensible artifacts, incomparably more convincing and ominous than any flying saucer, this was a hive mind, a collective whose individual components had all the flexibility of a human being, but were scarily enhanced as an interacting whole. In the TV show and the movies, humans eventually triumphed. But anyone who rationally assessed the stories realized that the Borg would really have won the day, leaving the Federation assimilated. Frankly, Captain Picard, transformed to "Locutus of Borg," came across better than he did as a human.

Although the main scientific thrust of the Borg as a creation is the combination of human and machine, it is worth briefly contemplating that "hive mind" aspect, because the more we find out about collective insects like bees or termites, the more impressively different and alien they are—an ideal model for science fiction. *Star Trek* was by no means the first to play on the idea that aliens may not be distinct individuals as we are, but part of a larger whole. In H. G. Wells's primarily comic novel, *The First Men in the Moon,* the Selenites form some of the most striking early alien depictions, and they are, in effect, a hive species.

Historically we have tended to think of bees or termites as lowly organisms, despite their remarkable construction abilities. But what was missed was the way that each insect is not a true individual. Instead it makes up a part of a larger whole, known in biological circles as a "superorganism." Just as the cells that make up our bodies are not individually conscious or capable of action, but the whole, combining the cells, is far more than the sum of the parts. Similarly, a superorganism combines the abilities of the different functional types of insect within the species, with specialisms to provide particular roles within the superorganism, and a relative lack of interest in the survival of any single insect, any more than we worry hugely if we lose a few cells from a finger.

In the Borg, that collective capability is combined with intelligence supplied by each component part. The cells that make up the superorganism that is a hive of bees or a mound of termites is capable of far more than each individual insect can do, so in the Borg, the combined superorganism is genuinely and scarily more advanced than any single human (hence the assertion that they would, without doubt, have beat the Federation). What's more, the linking mechanism of the Borg is much more flexible than insects can achieve.

Bees are primarily linked into their superorganism by the use of chemicals that they deposit and detect, and by physical movement, like the famous "waggle dance." This is sufficient to enable them to share information about sources of nectar, to construct complex and elegant structures, to provide air-conditioning to the hive by fanning with their wings, and to defend the queen under attack. The Borg, by comparison, have continuous broadband connection that does not require close physical proximity. This not only makes for much quicker action, but also more intelligent possibilities, reflecting the way that each component of the superorganism has human-like mental capabilities, not those of a tiny insect.

Impressive though the collective nature of the Borg is, it is their cyborg nature, bringing together living flesh, electronics, and mechanics that both fascinates and horrifies us in equal measures. The term "cyborg" was first proposed by scientists Manfred E. Clynes and Nathan S. Kline in 1960, and made visible to the public in *The New York Times* in the May of that year. Kline primarily worked with mind-altering drugs, and the initial definition that was used in the *Times* reflected this, saying that a cyborg was a "man-machine system in which the control mechanisms of the human portion are modified externally by drugs or regulatory devices so that the being can live in an environment different from the normal one." But the term has come to specifically refer to a melding of human and mechanical/electronic construct.

Real upgrades to the human body easily predate science fiction stories. Using a walking stick, for instance, goes back well into prehistory—and yet even this is using a device as if it were part of the body to enhance basic capability. Eyeglasses (sometimes attributed to the medieval friar Roger Bacon, though there is no good evidence that this is justified) were developed no later than the Middle Ages. And variants of the wheelchair have been around as long as we have had wheeled vehicles. Prosthetics too, replacing lost limbs, have been common as long as humans have survived such traumatic injuries. Wood and metal have been in use for at least a couple of thousand years (there is an example of an artificial leg in an Italian tomb dating back to 300 BC), though of course many fewer patients survived amputation in the early days without anesthetic or antiseptic.

Such simple attempts to replace functionality don't seem to trigger our emotions as something that is different and potentially scary. We don't consider someone wearing contact lenses or a basic prosthetic to be a cyborg, despite the person's life being enhanced by the technology. Where the difference becomes significant—and hence becomes interesting to science fiction—is where the modifications give a human abilities that go beyond those a normal human being is capable of, or where the technology becomes visually obtrusive, changing someone's appearance from that of our expectations for a "normal" human being to something that it is easy to regard as grotesque.

One of the earliest of the fictional cyborgs with extended abilities seems quite remarkable as a concept even today. This was in *The Clockwork Man,* a novel by Edwin V. Odle from 1923, which features a human being with a clockwork mechanism inserted into his brain. That mechanical bolt-on was used both to regulate his body and to enable him to (somehow) travel into extra dimensions. But a more familiar example of a cyborg with enhanced abilities while still being visually human was that critically injured astronaut Steve Austin, played by Lee Majors in the 1970s TV show *The Six Million Dollar Man.*

Based on the 1972 novel *Cyborg* by Martin Caidin, *The Six Million Dollar Man,* with its much-quoted tagline "we can rebuild him . . . we have the technology" took an astronaut who had suffered wide-ranging injuries and replaced his legs, one arm, and an eye to produce someone apparently human, but capable of superhuman feats of strength and vision. The show was successful enough to have a sequel in the arguably slightly more sophisticated *The Bionic Woman,* played by Lindsay Wagner. As is often the case with TV and movie science fiction, the physics of Austin's new abilities was laughable. It didn't seem to occur to anyone that when he lifted something incredibly heavy, even if his arm had super strength, the body it was attached to didn't, and so he would either collapse or the arm would be ripped off. But there is no questioning Steve Austin's cyborg credentials.

In the real world, bodily enhancements of this kind, usually after injuries, have rarely improved on the original, at least if the original is in the best of health, but have attempted to restore a semblance of normality. I have a metal plate in my shoulder as a result of a bad break. This doesn't give me any superpowers, apart from the not particularly desirable ability of being able to set off airport metal detectors, but it did restore the use of my shoulder. The same goes for those with hip and knee replacements or pacemakers—the reality of medical enhancements has tended not be an attempt to improve on nature, but to restore the status quo.

There is an interesting reflection on how the "visually obtrusive" aspect of cyborg technology has an effect on our perception in the difference between the way we react to cochlear implants and their visual equivalent. Cochlear implants help with a particular hearing problem. Sounds coming into the ear vibrate the eardrum, which passes the movement through three tiny bones (the smallest bones in the body) onto a membrane called the oval window, which sets the fluid in the cochlea in motion. The cochlea itself is a spiral-shaped bone chamber, which contains thousands of tiny tufts, which look a little like hairs, but are actually extensions of a cell membrane. As the fluid moves, it stimulates the base of these "hair cells" and generates signals in the auditory nerve. If the hair cells are damaged, hearing loss occurs.

Cochlear implants get around this failure by directly stimulating the neurons that the hair cells should act on. An external headset picks up sounds and processes the signal to produce a series of electrical impulses, which are passed by induction to a small device implanted under the skin. This then stimulates electrodes that are embedded in the cochlea. The earliest implants only had a single electrode, though numbers have increased over time to more than twenty separate stimulation points. Even so, this means that only a small subset of the hair cell connected nerves are activated, and it was originally assumed that the implants would just give a little guidance to help with lipreading, In fact, this relatively small repair has been enough to enable users to understand speech, proving much more effective than was originally expected. Over 300,000 people have now benefited from cochlear implants.

Although a cochlear implant is a true example of cyborg technology—it involves an electronic device implanted in the body with a direct connection to the nervous system—we tend not think of it as such, because it is no more visually obvious than a hearing aid. Take it a step further to the optical equivalent and it begins to seem much more like a true cyborg addition. The self-funded researcher William Dobelle made his first visual-aid brain implant in 1978, giving a man named Jerry a very limited approximation to sight using TV cameras mounted on the frame of his glasses, generating signals that were sent via a mainframe computer to electrodes implanted in his visual cortex. Over the years, Dobelle has refined his technology to a more portable system.

There is some controversy over Dobelle's experiments, as he works outside the academic framework and charges big money for the treatment. Working alone, he lacks the constructive synergy that groups in universities have through publication and teamwork. Even so, whether or not Dobelle's specific devices will ever become commonplace in the medical world, he has proved that an equivalent to a cochlear implant for eyesight is not impossible. His system does not restore sight in the normal sense, but gives an approximation to sight that is sufficiently good for basic awareness of the surroundings.

When the electrodes in Dobelle's device stimulate neurons in the visual cortex, the result is the production of phosphenes, small bright areas of apparent vision. We've all experienced phosphenes when we rub our eyes and little flashes of light appear to float in space before us. Here, the optical nerve is being stimulated by pressure. With Dobelle's implants, electrical impulses provide the stimulation. The initial calibration of his system involves putting a series of signals through the electrodes. The subject describes what he or she "sees," and the researcher builds up a map of these illuminations, like the pixels on a computer screen, except the phosphenes are much less regular in shape.

Once a complete map has been built, Dobelle's software can make use of the phosphene "pixels" to display images from the video camera that the subject wears. In the most recent version of his technology, there are two cameras, accessing two sets of electrodes on the left and right side of the brain, just as our optic nerves cross over to the opposite sides of the brain. Even with this latest version, the effect is limited, and at least one subject has had a seizure as a result of excessive stimulus to the brain, yet it does make it possible for the wearer to identify objects and to manipulate them.

Dobelle is not alone in this work—a number of universities have projects underway, and there is no doubt that there will be visual implants available much more widely within ten or twenty years. Some of the experiments go even deeper into in-brain electrodes, others are looking at stimulating the optic nerve externally, holding out the hope for some blind people (depending on the cause of the blindness) to have restoration of elements of sight without the need for intrusive brain surgery, just as cochlear implants can restore hearing without putting electrodes into the brain itself.

Once it is possible to feed the senses electronically, particularly if it is possible to do so without implanting electrodes, there is no need to stop at the limits of normal human capabilities. We can only see a tiny fraction of the full spectrum of light. Our eyes are limited to reacting to wavelengths between around 400 and 700 nanometers (a nanometer is a billionth of a meter). Extending below with longer wavelengths is infrared, microwaves, and radio. Above comes ultraviolet, X-rays, and gamma rays. With the right kind of camera, the same neuroprosthetic that enables basic sight could see the heat pattern of people in the dark, providing the equivalent of using night vision goggles, but linking directly to the brain.

Similarly, our hearing cuts off at a relatively low frequency—with an appropriate microphone, a cochlear implant could pick up the echo-locating squeaks of bats. As well as being able to see in the dark, these technologies coupled with appropriate sensors, would even enable us to stray beyond traditional sensory boundaries. It has been suggested that a dog's sense of smell is far more tightly linked to its eyesight than ours, making smells almost like an extra color that allows the animal to navigate by odor. Although it is natural to link light input to the eyes, we could equally "see" sounds from an electronic ear, or smells picked up by an electronic nose—it is entirely possible that a future sensory neuroprosthetic could give the wearer a much richer interaction with their environment than we are naturally capable of with normal human senses alone.

Similarly, the assumption tends to be that the pickups for the senses will be located in the natural place for them to reside—on the user's head. But with modern electronic communications there is no need for this to be the case. The sensory inputs could be coming from a tank on a battlefield the other side of the globe—or from a drivable probe on the surface of Mars. (The only trouble here is the time delay. Light, our fastest means of sending a signal, takes an average of four minutes to reach us from Mars, and another four minutes is taken up in getting a signal back, so the driver would need plenty of time to react to a problem.) And that's only the start. The signals entering the brain needn't even respond to a real sensory input at all.

Where watching a movie is always a more restricted experience than being out in the real world, the users of the brain-linked equivalent of a movie would see no difference between an electronic source and true vision. What's more, the image "seen" need not ever have existed in reality. As we saw in chapter 2, it could be the virtual world of a computer game, or the unreal universe that is formed by the World Wide Web. If the quality of images can be brought anywhere near that experienced in true vision, and particularly if it becomes possible to interface with the brain without surgery and electrodes (see page 22), this technology could transform how we interact with both the real and the virtual world, erasing the boundary between cyborg technology and Matrix-like mental worlds.

In 2006, two papers published in the journal *Nature* gave hope for a much wider use of neuroprosthetics to help those with damage to the nervous system regain the ability to interact with their environment. In the first, John Donoghue and colleagues at Brown University described how they implanted an array of ninety-six electrodes into the precentral gyrus area of the primary motor cortex, the part of the brain responsible for movement. The experimental subject, Matt Nagle, was a man whose spinal cord had been entirely severed in an accident leaving him with no control over his limbs.

Even though the experiment took place three years after Nagle's injury, he was able to "think" hand movement and produce signals through the electrodes that were able to move a cursor on a screen, to control simulated e-mail, and to operate connected devices such as a TV. He was able to do this while talking, as naturally as someone without his injury could talk and move their hand at the same time. He was also able to open and close a prosthetic hand and make basic movements with a robotic arm.

It might be imagined that this took a huge amount of training, gradually working up the tiniest elements of control to eventually achieve something usable (it's certainly how Hollywood would show it, probably in a montage with dramatic music in the background), but one of the surprises for the researchers was just how quickly Nagle gained control. Within minutes of being asked to imagine using his hand to move a cursor on the screen, he had succeeded, and then, without further training, he was immediately able to move on to perform the other tasks mentioned with a degree of success.

Although the experiment involving Matt Nagle took place at Brown University, it was undertaken in cooperation with the biotechnology company Cyberkinetics, and already has a trade name BrainGate, demonstrating the commercial plans of those involved. Even so, it clearly would be desirable to use a noninvasive connection. We have no experience to suggest how such fine electrodes would stand up to lifetime use, and the accompanying risk of infection or injury involved in the insertion operation make it highly desirable to work toward an external wearable connection. What's more, there is some evidence that for unknown reasons the response recorded by direct electrodes drops off after time, and different individuals seem to have different capabilities in using them (this may just reflect subtly different placement of electrodes).

The second paper in *Nature* demonstrates the ability to go beneath the concept of mentally moving a hand to the intentions behind it. Krishna Shenoy's group at Stanford University used a new approach with a series of electrodes in monkey brains to capture the intention of motion much more precisely. By using signals from the premotor cortex, they were able to predict the position of spatial targets the monkey was aiming for. This meant a significant speeding up of the link from brain to action. The researchers suggest the accuracy and speed provided mean that it would be possible to mentally control typing at a speed of around fifteen words per minute, if such a system were available to a human brain.

The significance of this development is that where someone using the BrainGate approach would have to consciously control an invisible hand—fine to move a cursor, but very slow for an action like typing—this approach would make it possible to think about typing and to have the appropriate signals produced, rather than consciously moving a virtual finger from key to key.

These experiments show very early steps toward being able to directly control robotic limbs from the brain. But to truly replace a physical limb there also needs to be appropriate feedback. We don't just need to be able to move a limb, it's also necessary to know where it is, an ability that's not as obvious as it seems. This requirement is essential for direct limb replacement. We have an awareness of the location of the parts of our body that is needed for fine control. It's this feedback that enables us to touch our noses with our eyes closed or, more practically, to direct our body to take action without consciously watching exactly what a finger or leg is doing. There is research going on in a few labs to try to close the loop and supply feedback to the brain, providing the kind of information that the nervous system provides from an arm or a leg.

We have now reached the stage that, though prosthetics with brain interfaces are still experimental, they have advanced significantly, often making less risky connections to nerves and muscles, rather than directly into the brain. As well as involving safer surgery, this seems to allow for longer effective use. For example, in 2014, research was published on a Swedish truck driver who had spent a year with a prosthetic arm that has a fixture to his bone and seven electrodes that connect into the muscle and nervous system in his remaining upper arm. The result has been remarkably precise control, including the ability to use keys or to pick up an egg.

Although there have been prosthetics for some time that use electrodes on the surface, or socket-type connections, these have caused a number of problems, not always working, producing sores and limiting precision. In the new experiment, the first to use implanted electrodes outside the lab, the approach has proved significantly more effective. Because it is using the same nerve links to neural pathways and muscles as the original arm, it simplifies both the connection and maximizes the amount of control available, but avoids the dangers of brain surgery presented by the early trials.

Of course, science fiction, lacking the practical limitations of dealing with the frail and immensely complex human body, takes the construction of cyborgs far further. The Borg work on the brain itself and take the extreme measure of replacing fully functioning body parts like eyes with "better" electronic equivalents. This, for most of us, is a change too far and evokes the reflex revulsion we feel for a creation like the Borg. As far back as the 1920s, science fiction has featured the ultimate version of this approach where a human brain is removed and given a totally mechanical body, whether a robot-like exterior, or providing the intelligence for an autonomous spaceship. Perhaps thankfully, medical technology is still a long way from being capable of producing such a cyborg.

In reality, the closest we have come to reproducing the sheer alien-ness of most movie cyborgs—does not involve humans at all. In 2006, DARPA, the U.S. Defense Advanced Research Projects Agency, put out a call for creators of insect cyborgs. It's an intriguing prospect (if you aren't too squeamish about the rights of insects). Our technology can do amazing things, but we can't yet come close to the way that nature manages to cram so much functionality in such a small package. After all, insects, some so small they can hardly be seen, can fly around with remarkable precision, all in a tiny organism that includes enough fuel to keep them going for hours. The closest we can get, model drones and helicopters, are crude and clumsy by comparison.

The kind of thing DARPA had in mind is summed up well by Michel Maharbiz's amazing, remote-controlled flying beetle. Maharbiz and his team at the University of California strapped a small electronic device to the back of a giant flower beetle. The beetle's electronic kit consists of a processor, receiver, and battery, with six electrodes wired into the beetle's optic lobes and flight muscles. This enables the scientists to send impulses into its nervous system at will. The optic lobe signals can be used to stop and start the beetle, while the flight muscle electrodes give directional control—the beetle's built-in systems handle all the complexities of keeping it from crashing, making the control far easier than it is with a drone or model helicopter.

As its name suggests, this is a large beetle, which gives it a good carrying ability, although its flight capabilities are not comparable to those of a fly. Eventually miniaturization could produce control rigs for much smaller insects where the role is surveillance. Imagine a security agency able to control a fly to land nearby a group discussing a terrorist activity—literally fly-on-the-wall monitoring. For other uses, the large carrying capacity of insects like the giant flower beetle could still prove valuable. Its ability to manage heavier payloads could make it not only a remote-controlled listening post but capable of delivering small physical objects. And though in the experiment the beetle was always in sight, with a camera on board—which can now be little more than pinhead size—the beetle could be flown like a drone, remotely from anywhere in the world.

Although not stretching to flight, anyone can now experiment with remote-control insects, thanks to a company called Backyard Brains, which provides the technology for a commercial, kit-form, cyborg cockroach. For under $100 you can purchase a "Roboroach"—the kit necessary to attach to a cockroach (insect not provided) and to control it remotely from a smartphone. Like the Maharbiz beetle, electrodes are inserted into the cockroach, in this case into its antennae, which gives the ability to provide left and right steering via the tiny backpack. (Backyard Brains will also sell you a box of cockroaches, if you don't have your own supply.)

It's easy to make fun of this bizarre business venture, but the fact that you can buy a remote controller for a cockroach off the shelf emphasizes that in simple form this technology is already maturing. The company argues that the Roboroach is an educational tool for understanding more about both neuroscience and cybernetics. While it's possible to question the ethics of the system used purely for entertainment, it is arguable that the opportunity for hands-on experience of the technology (and learning about the nervous system) is sufficiently valuable in an educational setting to make it acceptable.

It doesn't take a lot of working out to go from the idea of a brain-to-computer link (for humans or cockroaches) to thinking that that the computer could then be connected to another computer, anywhere in the world, via the Internet, that links to another brain, providing a form of cyber-telepathy. A team at Vanderbilt University under Jon Kaas has investigated direct brain-to-brain communication. By implanting electrodes in the brains of macaque monkeys, in the section of the brain responsible for hearing, Kaas and his team have been recording the signals produced when the monkeys pick up acoustic signals.

With a good map between what is heard and the stimulation that is received in the brain, the hope is that it will eventually be possible to produce the effects of hearing those sounds by replaying the stimuli into the brain. With similar studies on the brain cells that fire when sounds are generated, it should in principle be possible for two individuals to communicate brain-to-brain via an electronic link. Not exactly telepathy, but something very close (see page 223 for a crude alternative version of this that has already been demonstrated).

An $800,000 grant to Kaas is only part of the investment that DARPA has made in electronic interfaces to the brain. It has invested millions in another projects, all aimed at providing a better understanding of how we can interact directly with the brain, with the long-term goal of being able to control and get feedback from military craft, either to improve response times, or to remove the human to safety entirely, controlling the craft remotely, but with just as good a sensory input as if the pilot or driver were present in the hostile environment.

Many of the attempts to link the brain to electronics do so at the level of the brain itself, but others use our body's natural communication network to the brain, the nervous system, just as the prosthetic arm described on page 218 does. After all, this is how the brain habitually receives input and translates its wishes into actions, so it seems a sensible way to link into the brain. What's more, there's the advantage that the surgical requirements are less risky when, for example, connecting a chip to the nervous system in the arm, rather than by drilling holes in the head.

Since the late 1990s, Professor Kevin Warwick of Reading University in England has been experimenting with a range of implants under his skin that have been described as specific attempts to turn himself into a cyborg. Some of this is press overexcitement, but Warwick seems to revel in the press attention and has explicitly labeled himself, when implanted, as a cyborg. Admittedly, Warwick's description of himself is technically correct, but most users of the term tend to imagine a much more radical electronic/mechanical component to the mix.

Warwick's first implant was little more than the type of RFID chip now used to monitor inventories from bookstores to clothing and to accept contactless payment by tapping a card on a reader. On the upper inside of his left arm, a small chip in a cylindrical container was implanted between his skin and the muscle. For nine days it enabled Warwick to be recognized by various electronic devices around the offices of the department of cybernetics where he works.

As Warwick walked around the building, computerized voice boxes would say "Hello," doors would open, and lights were switched on and off by his presence. Of course, this was doing nothing that isn't possible less painfully with a simple security tag in the pocket or an ID badge, but the significance was that it demonstrated remote communication with an embedded chip. Warwick was to go significantly further in 2002. A second chip was implanted, this time with a 100-electrode array that was connected to Warwick's median nerve fibers, below the elbow joint of his left arm. With this connection, Warwick was able to have some control over both an electric wheelchair and an artificial hand, an early precursor of the 2014 Swedish development described on page 218.

Perhaps most interesting of all, Warwick's wife, Irena was also given an implant (though admittedly less complex). With this, she was able to send an artificial sensation to Warwick. A command from her brain activated her implant, which generated a signal. This was then translated into a signal sent to Warwick's implant, which finally generated a sensation in Warwick's brain. A communication from Irena's brain was sent to Warwick's brain by electronically extending and connecting their nervous systems. Warwick's implant was also stimulated to give a sensation by input from an ultrasonic sensor—in effect, he was "hearing" ultrasonics—and he was able to extend his nervous system over thousands of miles, connecting through the Internet to control a robot arm the other side of the Atlantic.

Warwick is still technically active in the field, but seems to have given up on the cyborg journey since the 2002 experiment, apart from writing a book on his experiences. He has frequently inspired controversy because he is very media savvy, and exposes his work more directly to media scrutiny than most scientists (though at the time of writing little has been heard from him in a couple of years). He may be a maverick, but he is not a charlatan. Certainly his work to date has been more "proof of concept" than anything practical and he has not been backward at getting his name in the press. Yet his work does offer significant advantages of relative safety over direct brain cybernetic connections, and there are some real hopes from his demonstrations that electronic extension of some aspects of human capability could be performed this way.

The latest in cyber-telepathy has come from Harvard University. In 2013 a human-rat link was set up via a computer using EEG (electroencephalograph) caps that monitor brain activity, enabling a human to move a rat's tail at will. But more interesting by far were results announced in September 2014, when Harvard scientists managed to communicate words directly brain-to-brain across a distance of 5,000 miles. The sender produced a message in binary by envisaging movements of their hands (a 1) or feet (a 0) while wearing a cap that detected activity in the part of the brain used for motor control.

The signal produced was sent through the Internet between India and France, and at the receiving end used a cap that used TMS (transcranial magnetic stimulation) to activate the optic nerve and produce flashes of light in the peripheral vision. It's not exactly cyber-telepathy as we know it, but it was dramatic both in not requiring intrusive and potentially risky brain surgery to implant electrodes and in the ability to communicate information this way.

A lot of cyborg technology, whether in the fictional Borg or these real-life experiments seems to allow us to communicate with others better—to be more aware of people and objects around us. But the technology of science fiction also can provide the opposite goal. It can make it harder to communicate and make it possible to hide people and objects away. Invisibility has long been a science fiction dream.

16.

ENGAGE CLOAKING DEVICE

Star Trek may have brought the term "cloaking device" into general usage, but I was aware of the concept of invisibility long before, thanks to H. G. Wells and a tradition of magical invisibility that goes back over 2,000 years.

The Time Machine, with its heavyweight political ramifications, bored me as a young reader, while *The First Men in the Moon* entertained and *The War of the Worlds* gripped, but the short novel *The Invisible Man,* that adds to Wells's impressive collection of science themes, horrified. The natural response when I was young to the thought of being invisible was that it would be fun. There are plenty of teen movies (or adult movies with a teen sense of humor) that take this viewpoint. I assumed, as the protagonists of those movies do, that invisibility would be a chance to be naughty and run around giving people shocks with unexpected pinches and dousing them with water—or spying on things we aren't meant to see—but for Wells, invisibility was not so much a gift as a nightmare for his protagonist, only ever known by his surname Griffin.

Griffin arrives at a country pub tightly clothed from head to toe. Any flesh that might be on view is revealed to be covered in bandages, his eyes concealed behind goggles, his nose a shiny fake. We soon discover that Griffin is a scientist from University College in London, a medical student who has studied physics, particularly the manipulation of light, and has produced by chemical means permanent invisibility. Whether due to the realization of his fate or the effect of the chemicals, he descends into malicious madness, killing a hapless old man and terrorizing the neighborhood before he is finally battered to death. Griffin's occasional assistant is a tramp called Marvel, which has a certain irony when you consider how similar Griffin's tortured response to his "superpower" is to the way that Marvel Comics' heroes tend to agonize over their abilities.

When Griffin is given the chance to explain his invisibility, he points out that the only reason we see objects is because they reflect and absorb light. So, when the white light from the Sun hits a red fire hydrant, the paint absorbs the light, but re-emits the red photons, so we see the hydrant as being red. (This is usually described as reflection, though in practice all the light is first absorbed, it doesn't bounce, like a tennis ball off a wall.) Griffin suggests (incorrectly as it happens) that almost all human cells are transparent: "[I]n fact the whole fabric of a man except the red of his blood and the black pigment of hair, are all made up of transparent, colourless tissue." Griffin's experimental chemical treatment renders his blood cells transparent, and making use of special "ethereal vibrations" (Wells hooks in to the enthusiasm for the newly discovered X-rays by having Griffin comment "No, not those Röntgen vibrations") he alters his body's refractive index to be the same as air, so that (unlike glass) it is not still visible despite being transparent.

Although the mechanism itself is meaningless, this idea of having the same refractive index making an object invisible is indeed true. While transparent glass beads can still clearly be seen in water, as they have a significantly different refractive index, there is a clear plastic with a refractive index that is very near to that of water, and when beads of this material are dropped in, they seem to disappear.

What is never explained by Wells is a problem that almost all invisibility fiction ignores—particularly the movies—how Griffin was able to see. If a person has been rendered fully transparent—or for that matter, if some sort of cloak is diverting light around the person as appears to be the case in the Harry Potter films or a *Star Trek* cloaking device—then light from the surroundings is not interacting with the invisible being's retinas. We only see because the retinas at the back of our eyes absorb photons of light, triggering electrical signals that are interpreted by the brain as sight. If that is happening, then light that hits the retinas won't come out the other side. You would see the retinas hanging in space. The fully transparent invisible man would also be totally blind. Similarly, a cloaked warbird in *Star Trek* should not be able to see anything in the space around it.

As this and the book's subtitle "*A Grotesque Romance*" makes clear, although Wells dressed up his story with the trappings of science *The Invisible Man* was far more a continuation of the way that invisibility had been rendered in myth and fantasy, ever since the stories of Gyges, an Ancient Greek king who took the throne with the aid of a ring which enabled him to become invisible. Suggestive of the much later and far darker ring of J. R. R. Tolkien, the Gyges version did not make the wearer invisible all the time, but was operated by twisting it around the finger so that its decoration faced inward. Almost always in fiction, the ability for an individual to become invisible corrupts that individual—and the temptations that invisibility offers prove to be largely criminal. The invisible person is not inclined to perform acts of kindness and valor, or even to hide from enemies. Time after time, he or she will peek and pry where invisible hands should not reach, will steal or abuse others in misdemeanors from petty pinching, eventually pushing to sexual transgressions and murder.

In early stories, the ability to become invisible was usually produced by something that was worn—a ring, a hat, or in the mode that made its way all the way through to Harry Potter (and that verbally, at least, features in *Star Trek*), a cloak. The cloak, at least, provides a mental image of something that covers the body and could hide it away if it only had special properties, where a mechanism like a ring is much more clearly a focus for magic. In medieval times, these fairy-tale sources of invisibility, that offered no mechanism but simply did the job to make the point, were replaced by magic spells that often involved complex rituals and unpleasant ingredients. If the character was lucky they might only need to find everyday requirements like a handful of beans, but more often than not, the head of a corpse or the boiled bones of a cat would also be required.

The science writer Philip Ball has suggested that the lengthy rites portrayed in "real" books of magic, involving near-impossible-to-obtain items and to-the-second precision on the position of the Moon or planets indicated that these magic spells were never intended to be used. Instead, he suggest, they were the equivalent of a Mason's secret handshakes and rituals, an entry by knowledge as an adept into an inner circle of a secret society. There may be some truth in this, though it seems likely that the complexity of the rituals were also, and perhaps more fundamentally, an example of the classic hustler's get-out clause. If you make sure that the spells in your magic book are impossibly complex and difficult to perform, when the customers come back to complain that they don't work and ask for their money back, you can always find an excuse—they didn't undertake the ritual at the exact stroke of midnight, or it was the wrong kind of cat—to show why nothing happened.

In nature, the ability to become at least partially invisible seems more often to be about protection than an opportunity to cause mischief. Of course it is helpful for any predator to approach stealthily, and often they will make use of their surroundings to help conceal themselves, whether it is a big cat stalking through the long grass or a trap-door spider hidden behind its well-concealed flap of soil, vegetable matter, and spider silk. This example of the concealed aggressor is reflected in science fiction in a whole host of invisible attackers preying on humans, such as the alien in the Predator movies, a theme that came originally out of fantasy and horror. To the Victorian mind, when these stories were often first written, an invisible threat could be made more terrifying than anything that could be described. But in nature it is more likely to be a prey animal than a predator that evolution has shifted toward markings, or even active coloring that will enable it to disappear as much as is possible against its background.

Perhaps the best-known user of protective coloring in nature is the chameleon, which is ironic as this odd-looking lizard really isn't very good at camouflage. Some (but not all) species of chameleon have special cells in their skin called chromatophores, which contain different layers of pigments that can be distributed evenly or collected together, changing the coloration of this "skin pixel." But chameleons are limited in their ability to control this facility and don't seem to use it primarily to conceal themselves, but for a combination of heat management (making their skin darker when it is cool and lighter when it is hot) and as a means of signaling to other members of its species.

The majority of animals that do go in for concealment have a less complex approach than chromatophores. Prey animals that have permanent markings that blend better into their surroundings tend to be more likely to survive, so evolutionary pressures encourage keeping any such markings that appear as a result of the random forces of evolution. This process has been observed happening with remarkable speed in the peppered moth. These common moths have a pattern on their wings that helps them blend in well on a lichen-covered tree, so that they fade almost invisibly into the background and are very difficult to spot when unmoving on a tree trunk. (As always, such camouflage is useless in other circumstances—the moths are all too visible against a brick wall or a green leaf.)

As the Industrial Revolution bought factories that churned out air pollution on unprecedented scales, the lichen was largely killed off by pollutants, exposing darker tree bark that was made darker still by the soot that was deposited on it. The result was that the naturally darker variants of the peppered moth were less easily spotted by predators and in just tens of years, natural selection made it more likely that these darker moths would survive and breed, so the peppered moth population grew darker and once more were hard to pick out against the tree bark. Since then, things have changed. In the highly industrialized areas where the dark moths developed, clean air legislation has meant that the trend has reversed. Bark has become lighter as lichen regrows and light-patterned moths once more dominate, able to conceal themselves on lichen-covered trees one more.

This "industrial melanism" makes an excellent example of evolution in action (and emphasizes that evolution need take place over a long period of time if there are strong drivers and relatively short life cycles). The use of such markings to give a degree of concealment is common across a far wider range of organisms than moths alone. But when animals use this approach, they face the same problem as experts given the task of trying to conceal warships, soldiers, or military aircraft. A pattern that may be very effective at concealing a zebra or giraffe, say, among a broken up backdrop of light and shade like a jungle is hopeless at hiding it when it is seen against the sky in an open savannah. Similarly, camouflage markings on a vehicle that work well in a desert would be useless on a background of green hills. Having markings that match a particular background, or that break up an image to make it easier to hide in mottled shade, can be a disadvantage in all but a limited set of circumstances. And those seeking to build a real cloak of invisibility have had to instead consider a more active natural model.

In *Star Trek*, we first met the possibility cloaking in "Balance of Terror," a 1966 episode that had intentional echoes of a classic movie subject, the pursuit of a submarine. The cloaking device (given this name a couple of years later, when Kirk went on a mission to steal one) was used by the Romulans to conceal their "bird of prey" ships, which would ripple and disappear as the device was engaged. Although the devices would later also be used by the Klingons, the Federation seems to have considered using them to be cheating, and despite Kirk's successful mission, cloaking was banned by the Federation. Although Gene Roddenberry seems to have considered this a moral decision because "our heroes don't sneak around," in practice it is more likely to have been necessary to stop the story lines from being ruined. It would have been hard for there to have been much plot development if all ships could become undetectable at will.

There was never any significant attempt to give a scientific explanation of *Star Trek*'s cloaking technology, and while it could be said to inspire some of the more interesting ways that we can actually construct an invisibility shield, the most advanced approaches, in practical terms at the moment, are those that bear more resemblance to natural solutions than anything that is found in science fiction. (If there were a science fiction forerunner, it might be the light-bedecked alien ships of *Close Encounters of the Third Kind,* which seemed more designed for communication than concealment.) The stealth technology that we hear so much about in the military is primarily about making use of fairly simple physics and does not try to do anything so sophisticated. It involves minimizing the way that light or sound reflect off a plane or other stealth vehicle to make it appear that the vehicle is much smaller than it actually is. But just as the early designers of camouflage learned lessons from the patterns that gave animals concealment, so the existing approaches to true invisibility seem to be inspired by the remarkable natural capabilities of a flatfish.

A number of bottom-dwelling flatfish manage to do what chameleons are supposed to do, but the fish make it work properly. Perhaps we should call someone who blends into his or her surroundings a flatfish, not a chameleon, although the comparison doesn't sound as flattering. Like the chameleon, the flatfish has a variety of chromatophores on its skin, or to be precise its upper surface. These enable the fish to contol the range of colors displayed to anything seeing it on the seabed. So far, so very chameleon-like, but the flatfish has an extra piece of biological technology. It couples the color display surface on its upper side with a series of color-sensing patches on its underside. The result is that it can take a kind of biological photograph of the surface beneath, before it settles. The flatfish then transmits this pattern of coloring, light, and shade to its upper surface. The result is that, when it works well, it can blend with startling effectiveness into its surroundings.

When it comes to human-cloaking devices, taking a lesson from the flatfish seems our best bet at the moment. It is hoped that this same approach can be taken on a much larger scale when the proposed Tower Infinity is built on the outskirts of Seoul in South Korea. The 1,500-foot high tower will have video cameras, taking in the view around it at three levels. The data from these cameras will then be processed by computers to allow for varying viewing angles and fed to a vast number of LEDs set around the outside of the building. The whole building will then act like one of the vast LED screens used at events and stadiums, projecting the view from the other side of the building, so that it should disappear from view, or at least appear as little more than a fuzzy outline. There's a reasonable question of quite why anyone would want to make a major piece of architecture disappear, particularly one that is tall enough to be a hazard for low-flying aircraft—but the aspiration is still impressive.

The principle is clearly very similar to the flatfish—more so than the "invisibility cloaks" that often crop up in Internet videos, demonstrated for a number of years by engineer Susumu Tachi, working at Keio University in Yokohama, Japan. Tachi projects an image onto clothing captured from cameras behind an individual—but although the picture is made pretty well as bright as natural lighting (always a problem with projection) by coating the clothes with a layer of tiny, high reflection beads, the effect is a static one and is only visible from a particular direction. What's more, it requires a whole collection of external equipment instead of acting as a standalone cloak.

In fact there is an old magic illusion that dates back over a hundred years that can be significantly more effective than the hi-tech solution in this situation. A frameless mirror is placed in front of part of a person. They are located in surroundings that are pretty much the same in front and behind—for example in the middle of a large grass lawn surrounded by shrubs. The result is that the mirror reflects a view that is actually from behind the viewer that is pretty well identical to what you would expect would be behind the subject. If everything stays static and the viewing angle is carefully selected to avoid the viewer seeing his or her reflection, the effect is very much of seeing straight through the subject—though it only takes a person to walk past either behind the subject or behind the viewer, hence visible in the reflection, to spoil the effect.

In fact the whole problem of the way that views alter with changing directions and perspectives is the often-ignored devil in the detail of the kind of invisibility cloaks that rely on projecting an image of what is behind an object from the front of it. The flatfish gets away with these problems in part by being thin and flat, so there is little variation in the surface, and also because the image it provides is just a pattern like the seabed, not a true three-dimensional view with perspective. As soon as an invisibility cloak is required to reflect a real, complex, three-dimensional environment, the problems involved multiply rapidly.

Firstly, the cloak would ideally be viewable from any direction, so that you can walk around the concealed person or object and whatever is hidden wouldn't suddenly appear as you see around the corner of the shield. This is probably the least of the practicality issues, as with modern electronics it is certainly feasible in principle to have a vast number of tiny cameras interlaced with tiny LEDs so that each surface is a mix of cameras and screen, displaying the view from the opposite side. This approach has been proposed by Franco Zambonelli at the University of Modena, Italy, who suggests that while wiring up such a cloak may prove impossibly fiddly, we are very close to being able to produce the myriad cameras and screen elements that would be connected by a local wireless network, so they would not need wiring to connect them up.

More difficult still is that we are very sensitive to the way things change in appearance as we move around. The miniature display elements couldn't produce a really deceptive image simply by using the kind of straightforward LED used in a normal large screen (or as proposed for the Seoul skyscraper), because as soon as the viewer moved it would be clear that they were not seeing a normal view. To make the effect convincing, each tiny screen would have to be angled so that it gave subtly different views over up to a 180-degree field of vision. This is certainly beyond current technology, but it is not impossible to conceive such a multiangle viewing device.

These cloaks that conceal items by providing what is little more than an enhanced, short-range TV system are probably the most convincing possibility for cloaking in the short- to medium-term. They produce the best effects that we could come up with in the next few years. But the long-term hope is to make use of metamaterials to produce something far closer to a true cloaking device. These mind-boggling materials work by turning the familiar world of the refractive index on its head.

Whenever light moves from one medium to a denser one, it bends inward toward a perpendicular line into the material. The more light is slowed down by the material, the more it will bend. This has been called the "Baywatch principle," because like a lifeguard, the light takes the quickest route. We tend to think of a straight line as the quickest way to get between two points on a flat surface, but this is only true if we can achieve the same speed over the whole distance. When a lifeguard spots a person in trouble out at sea, the natural tendency might be to run straight toward the victim. But the clever lifeguard runs farther across the sand, so they can then cover a shorter distance in the sea—because however strong a swimmer they might be, they can travel faster on land.

One of light's strangest properties is that it appears to be lazy (or clever)—because it too inevitably takes the route that will take the least time. If light moves from air into water, it will travel farther through air, then bend at the point it hits the water to head more directly toward its destination, so it covers a greater distance in the air, where it can travel more quickly. This process, called refraction, is why pencils seem to bend when you put them in water, and how lenses work, as they are shaped so as to bend the light (as it travels from air to glass, then glass to air) to head toward a particular focal point. The bigger the refractive index of the second material compared to the first, the more light will bend in toward a perpendicular line into the material at the point of impact.

Over the last few decades, artificial materials called metamaterials have been constructed that have a *negative* refractive index. This means that instead of bending inward as it goes into a material where it is slower, the light bends outward. The result is that light can be manipulated in whole new ways and, notably, it can be sent around an object, bending outward around the object to be concealed and then bending back in, so that when an observer looks at an object protected by the metamaterial cloak, they see the light from behind it instead of the object itself. In principle, then, we have a true invisibility cloak—but even more so than with the TV camera and screen versions, there is a lot of technological development to be undertaken before we have something that could be deployed for real.

Firstly, with the current materials, this technology works best on microwaves. A metamaterial is often made up of a repeating pattern, and the scale that pattern has to be depends on the wavelength of light it influences. Microwaves, with wavelengths around a centimeter, are much easier to build metamaterials for than visible light, where the wavelength can be 100,000 times smaller, which makes for an impractically small construction using most techniques. Visible light metamaterials can be constructed using the same kind of technology as is used to produce the circuits of computer chips, but as yet do not allow for the construction of a practical invisibility shield.

Current shields also tend to only work in a single direction while concealing a very small object—to be of value, a true invisibility shield is liable to have to work from any direction and to be able to hide a sizable object. In the end, the problems that are faced are engineering restrictions instead of anything truly prohibitive in the physics, but true, large-scale, visible light invisibility shields are well beyond current construction techniques.

The best working forms of cloaking today makes use of a pragmatic view. Don't try to achieve perfection, but instead do what is possible, and do it in such a way that it delivers value. The engineers working on military concealment recognize that it is impossible to make something clearly and consistently disappear, and instead act to make it look like something else instead. As we have seen, stealth technology on aircraft aims to make the aircraft appear like a much smaller, probably natural object that radar operators would ignore, rather than making it disappear entirely. An even cleverer approach has been taken by defense technology company BAE Systems, making use of the relative ease of fooling the infrared technology that is widely used for night vision.

Usually a tank is easy prey for anyone using night-vision goggles, because when the tank's powerful engines are running, the metal hull becomes much hotter than its surroundings and it glows like a tank-shaped beacon when see through the goggles. BAE's Adaptiv system tackles this, but not by trying to make the tank disappear, which is pretty well impossible. Instead, the system heats and cools a series of hexagonal panels on the outside of the tank to transform the appearance of the vehicle. Look at a tank through infrared goggles when the system is turned on and instead of a large, distinctive tank, you will see a small family car, complete with wheels and windows. The actual shape of the vehicle totally disappears, and the apparently innocent car explains the inevitable detection of *something* out there, because absolute concealment is pretty well impossible, while making what is seen appear harmless.

The kind of invisibility that H. G. Wells imagined, where a human being totally disappears from sight due to changes in the structure of their cells, is far from being possible even today—and chances are, would not be desirable if it were. But some aspects of cloaking have more resemblance to the science fiction originals than do many of the scientific developments we have seen in this book.

Hiding something away is certainly a lot less complex, and a lot less challenging to our technology, than attempting to produce machine intelligence. Even though we now have phones that can outpace many of the basic computing functions envisaged in earlier science fiction, where computers were often assumed to be vast house-sized constructions, true artificial intelligence still seems far from our grasp.

17.

OPEN THE POD BAY DOORS

We remember very little from our preteen years as we get older, with many apparent memories reconstructed from photographs, videos, and family stories, but one memory I do have (admittedly from the grand old age of twelve) is going to see *2001: A Space Odyssey* when it first came out. We saw it in Cinerama, the short-lived ultra-widescreen format that was developed in an attempt to woo audiences back to the movies from the all-too-easy draw of the TV. Cinerama was, on the whole, a flop. The original process, requiring three 35mm cameras and three synchronized projectors was too expensive and impractical to gain a wide audience. *2001: A Space Odyssey* was shot in a budget version of the process, Ultra Panavision 70, which used a single 70mm camera, losing some of the dramatic wraparound effect but still requiring an expensively modified theater. Still, for those of us who got to see it, the result was a very wide 2.76:1 screen ratio, unlike a basic modern widescreen movie, which is more like 1.78:1.

For a young member of the audience like me, the initial "Dawn of Man" sequence of the film, which shows early hominids being influenced to use tools by the appearance of a mysterious black monolith, was very long and dull. But once we had the famous match cut from a flung bone to a spaceship, I was enthralled. Even today, inured as we are to visual wonders from Hollywood, watching *2001* back on the less-extensive screen of a TV it is impossible not to be impressed by the stunning graphics. Bearing in mind this was a pre-CGI world, the film's technical crew worked wonders—space looks real, down to the rarely repeated accuracy of all external shots being totally silent. And the rotating artificial-gravity environment inhabited by the *Discovery One* crew on their mission to Jupiter is stunningly rendered.

This was supposed to be a wheel-like structure that was rotated to give artificial gravity, so "down" to the astronauts was outward on the wheel. As the astronauts casually walk around this circular footway on the movie they go from being upside down (from our viewpoint) to correctly oriented, all achieved, remember, without CGI. In reality, such a feature wouldn't work. You can, indeed, accelerate things this way—and as Einstein pointed out, acceleration and the pull of gravity are indistinguishable. Many of us have been on carnival rides where the riders are rotated in a drum, then the floor falls away, but everyone sticks to the drum because, in effect, the rotation causes gravity. But such rides also make people feel queasy. Unless the wheel is big enough for rotations only to take place twice a minute or slower, motion sickness makes them impractical. And to provide Earth gravity, such a wheel would need to be around 500 meters (1,500 feet) in diameter. So the wheel in *2001* was too small—but the principle is good.

It was impressive that the rotating wheel could be made so realistic looking at a time when the effects couldn't be mocked up on a computer. Stanley Kubrick's studio at MGM Borehamwood in London made use of a huge rotating set. The actor, appeared to walk around a stationary footway, but in reality, the wheel was rotating and the actor stayed at the bottom, as if walking in a medieval treadmill or a hamster wheel. By moving the camera in synch with the wheel, it appeared that the surroundings were stationary and the actor was moving around inside. The mechanism could be a little risky: the crew had a tendency to leave items on the wheel, which would then crash to the floor as they were lifted high in the air. This was most dramatic when lightbulbs plunged and exploded, but came close to disaster when a large wrench flew through the air to miss a visitor to the set, AI scientist Marvin Minsky, by just a few feet. According to Minsky, "Kubrick was livid and quite shaken and fired a stagehand on the spot."

Of course, the spaceships were wonderful, so different from the crude, missile-like ships that science fiction movies had portrayed before. And the idea that the shuttles from Earth to the space station could be operated by Pan Am, while the video phone was run by Bell (that's the danger of taking existing brands into an imagined future) was intriguing. But the most amazing thing to me back then was Hal, the talking HAL 9000 mainframe computer responsible for running the *Discovery*. At this point in my life I had never *seen* a computer, other than on science shows on the TV, though very soon after I would be bought my own. It's hard to emphasize just how different Hal's human-like intelligence, unblinking electronic eye, and calm voice was from my Digi-Comp I. My computer was a mechanical device, worked by pulling a plastic plate in back and forth by hand.

The Digi-Comp I included three plastic plates that could be "programmed" by adding pegs, a little like chopped up drinking straws, to the side of each plate. These pegs caught on pivoted vertical wires that ran down the side of the computer. Depending on their position, the pegs would either hold the wires in place or flip the wires' positions as the device was cycled by pulling the main plastic lever in and out. The output was produced on three 0/1 mechanical binary digital displays. So with pegs in the right positions you could program the Digi-Comp to undertake logic operations like AND or NOT or NOR, simulating the gate mechanism that is present many millions of times over at the heart of every digital computer. It was the geekiest of geek toys and it took a whole lot of imagination to see it in the same light as a HAL 9000.

Although I had heard of Pan Am and Bell, at the time I had no awareness of the behemoth of 1960s computing that was IBM or of the possible link between the IT company and the movie's computer star. Pundits were quick to point out that not only was the naming convention "HAL 9000" reminiscent of IBM model names like the "IBM 7000," the first of IBM's transistorized computers introduced in the early 1960s, but also IBM could be transformed into HAL by simply moving each letter back one position in the alphabet. Both writer Arthur C. Clarke and director Stanley Kubrick, each with a major hand in every detail of *2001,* denied that this was intentional, but many think that the coincidence is just too unlikely to believe that there wasn't at least an unconscious leap from IBM to HAL in the minds of the writers. In some ways this transformation is less clumsy than the official origin of the name as an acronym for Heuristically programmed Algorithmic computer (for which HAC or HPAC would be better).

Clarke specifically commented that "I was embarrassed by the whole affair, and I felt that IBM, which was very helpful to Stanley Kubrick during the making of *2001,* would be annoyed." Clarke pointed out that the chance of selecting the letters randomly was 17,576 to 1 against, admitting this made his claim seem unlikely. In reality he didn't think about the probability very well as this figure produced by multiplying 26×26×26 is far too big. There are plenty of other combinations they could have chosen that would also suggest IBM, like JCN (the result of moving forward instead of the backward move that produces HAL), IBC, or JBM to suggest but three. In practice, they are likely to have wanted the computer's name to sound like a person's name, which if restricted to three characters leaves relatively few alternatives, like JIM, MAC, SAL, SAM, and TOM. So perhaps a coincidence isn't as unlikely as it seems.

Like many science fiction visions of the future, Hal manages to be both far too advanced and far too basic at the same time. While most of the technology demonstrated in *2001* still looks convincingly futuristic, the designers (and presumably IBM) were unable to grasp just how advanced computer graphics would become. The navigation computers on the flight deck of the lunar shuttle and the screen on which Hal plays chess with Frank Poole, have graphics more like those available on the early home computers of the 1980s—for the cognoscenti, the navigation screens were particularly reminiscent of the first wide-ranging space game, *Elite,* which was released in 1984—than the capabilities possible by the year 2001. The images look painfully crude. What's more, when Hal is eventually disabled, his computer room environment bears more resemblance to the room-sized racks of circuit boards that made up an old-fashioned mainframe than the distributed environment of modern networked devices.

On the other hand, the level of artificial intelligence displayed by Hal—the ability that the computer has to respond uncannily like a human being—is far beyond 2015 technology, let alone what was possible in 2001. Hal even manages impressive emotional responses. In fact, arguably, the computer displays far more emotion than all the other characters in the film, who perhaps intentionally were made relatively emotionless, providing a striking contrast. When AI pioneer Marvin Minsky was interviewed about Hal in 1997, he put the timescale for making this kind of technology possible "in between four and four hundred years." This is a much more honest and likely estimate than most attempts to predict the future of technology, even if the bottom end of that prediction was ridiculously optimistic and is long past.

A fascinating examination of a HAL 9000's thinking abilities is made by IBM research scientist and member of the Deep Blue team, Murray S. Campbell. Deep Blue was IBM's foremost chess-playing computer. Part of a series originally named Deep Thought after the super-powerful computer in *The Hitchhiker's Guide to the Galaxy,* Deep Blue was first constructed in 1995. Two years later, an upgraded version succeeded in beating reigning world chess champion Garry Kasparov across a series of six games (it had won a single game the year before, but lost the series). As we have seen, during the movie, Hal and crew member Frank Poole play a game of chess. For some reason, mechanical chess players have held a fascination for far longer than the technology has been possible to construct, resulting in a kind of real-world science fiction construct of the 1760s in the form of the Hungarian Baron Wolfgang von Kempelen's remarkable mechanical Turk.

This complex automaton consisted of a chessboard with the top half of a magnificently dressed humanoid figure attached, which would take on all comers and often beat them at chess. The mechanical Turk would "outlive" von Kempelen and ended up having a second career on the other side of the Atlantic in America. Its full lifetime, during which it had varying fortunes, spanned eighty years. Given the technology of the day, the Turk had to be an elaborate trick, and so it proved to be. The fancy cabinet that supported the chessboard housed a very small chess master, who manipulated the Turk's hands and directed the play by pulling on a series of levers.

The first realistic concepts that could lead toward an automated chess device came from the grandfathers and father of the modern computer respectively, Charles Babbage and Alan Turing. Babbage speculated that his mechanical programmable computer, the Analytical Engine, which was designed but never built, would be able to play chess, while Turing wrote a simple program for chess playing that was only ever executed by hand. More impetus was given by information theorist Claude Shannon in the 1950s. Shannon made use of John von Neumann's minimax algorithm, which would give a score to different possible moves and used it to calculate what was thought to be the optimum strategy. Shannon never produced a workable program, but by the time *2001* was filmed, there were crude chess-playing programs running on mainframe computers, and in 1973, David Slate and Larry Atkin wrote Chess 4.0, the first truly effective software that was able to make use of a computer's strengths and play a game that could beat most everyday players.

Today's computers are regularly able to beat grand masters; since Deep Blue beat Garry Kasparov this has not been in doubt. Chess programs that can run on a phone are capable of winning a game against casual players with ease. But the Deep Blue expert Campbell points out something that Kubrick and Clarke are unlikely to have considered in scripting the Hal versus Poole match in the movie. This is that computers and people play chess in very different ways. The approach taken by human experts in playing the game demonstrates a form of intelligence. They manage to combine knowledge of the basic rules, strategies, and past games with the ability to think through the move choices available, a kind of holistic approach unavailable to a computer.

By contrast, a dedicated chess-playing computer like Deep Blue does not make use of any real intelligence. Instead, such a computer has the ability to work through many billions of possible ways that the game could develop in an extremely short time, and is able to score those different scenarios to produce the move that the machine considers to be the best available at this point in the game. Deep Blue does not "understand" chess at all. It couldn't describe what a pawn is, or what makes a good move. Instead it has the ability to run through sequences of moves and to assess possible outcomes by giving them weightings. This might seem to make it unbeatable—as it certainly would be if it could run through *every* possible move that could be made. In practice, this isn't possible, as there are estimated to be more possible chess positions than atoms in the universe.

An expert like Campbell can tell by examining the chess game that takes place for thirty seconds or so in the movie what kind of player Hal appears to be. And the answer is that he plays like a human. This clearly isn't necessary just to beat Poole—any off-the-shelf software running on a perfectly dumb PC could probably do this, with no need for artificial intelligence. Yet Hal demonstrates a human-like approach to the game. Apparently the game that Kubrick used is a real one, played in Hamburg in 1913. In practice, of course, this was because Kubrick, something of a chess enthusiast, wanted to make sure that the game play was appropriately realistic.

Despite Campbell's claim, the style of play used doesn't prove Hal's true intelligence. In principle, the outcome could still be the result of one of the methods used by Deep Blue and the other grand master machines, which included a database containing hundreds of thousands of existing games. Although this particular game was not in Deep Blue's database, the chances are that a real Hal (especially one that had access to the Internet, something never dreamed of by Kubrick and Clarke) could have accessed the 1913 game and could have made use of it as a way of initiating play. There is no guarantee, of course, that Poole would have continued the same way that the losing player did in Hamburg all those years before, and in reality the game is most likely to have deviated—but the outcome shown is possible.

Simply playing the whole of an existing game is not a practical tactic, and isn't the approach taken by Deep Blue and other chess computers. As Campbell points out, the programs tend to assume that their opponent will make the best possible move. But this particular game involves a trap, where Poole is fooled into avoiding the best possible way of playing. Such traps are alien to computer play and it is this in particular that Campbell tells us suggests that Hal is genuinely playing with intelligence—or at the very least it shows that Hal's style of computer play makes much more significant use of existing games than typical chess programs to include the possibility of traps.

Although playing chess provides interesting clues to Hal's mental capabilities, it isn't in itself a particularly valuable skill for a general-purpose computer. A game of chess might be entertaining, but it doesn't keep the ship on course, or the crew healthy. A much more interesting human-like feature from a practical viewpoint is Hal's ability to converse with humans using perfectly accented speech. In the movie, of course, this was easy enough to do by having an actor read Hal's lines, but the realities of making a computer capable of holding a meaningful conversation have proved far tougher than most researchers thought would be the case back in the 1960s when *2001* was made.

There are two huge challenges and one small one. The easy part of the equation is getting Hal to speak in a fluent fashion. The halting robotic speech that has become the acoustic trademark of physics professor and media star Stephen Hawking is a very early piece of kit, which Hawking has kept because it has become his recognized voice, but the technology is now well advanced. Though it is unlikely that we would confuse a speech generator with a human, good modern text to speech systems can work through pretty well any piece of writing and make it comprehensible.

Even in the time I've had a satellite navigation system in my car, things have moved on hugely. The first generation tended to reproduce place-names phonetically. The second was much better, but still was caught out by strange English place-names like Cirencester (pronounced siren-sess-ter), and would refer to a road junction labeled 11A as "eleven ah." Now, the third generation handles all these effortlessly in a voice that is startlingly human-like. It's interesting that in the far less imaginative original series of *Star Trek,* which was produced around the same time as *2001,* computer speech was portrayed as being much more mechanical than Hal's smooth tones, despite the TV show being set over 250 years ahead of *2001.* In the real world, not long after the start of the twenty-first century, even a small GPS device can produce synthetic speech that can sometimes be mistaken for the real thing.

Talking inanimate objects have appeared in fantasy as far back as Homer, and were a regular diet of the medieval rumor mills, which attributed the ownership of oracular brass heads to a number of the early proto-scientists who were seen as potentially dangerous magicians, including the tenth-century French cleric Gerbert of Aurillac, and the thirteenth century pair of the Bavarian Albertus Magnus and the British Roger Bacon. In reality, Bacon particularly would have been horrified by this attribution of magic to someone who despised magicians as tricksters and attempted to take a scientific viewpoint on the world. But after his death a mythology built up around this remarkable friar to the extent that by the sixteenth century his fictional adventures were published as *The Famous Historie of Fryer Bacon*. In it, we read of the talking head he is supposed to have constructed:

> Fryer Bacon, reading one day of the many conquests of England, thought how he might keep it safe against future conquests, and so make himself famous for posterity. After great study he found that the best way to do this was to make a head of brass, and if he could make this head speak (and hear it when it spoke) then he might be able to wall all England about with brass.

Inevitably it took the real world a significant time to catch up with fantasy. It's a lot easier to tell a story about a talking head than to construct one. Remarkably, the first recorded attempt at a real mechanical speaking machine dates back to 1779. It was then that a C. G. Kratzenstein produced a complex apparatus that used bellows to blow air across a reed inside a chamber shaped to give similar resonance to the human larynx. This managed to produce only vowel sounds, but within twelve years a competitor, Wolfgang von Kempelen of mechanical Turk fame, had made what was almost a speaking musical instrument. Again a reed worked with a resonance chamber, but in this case the shape of the chamber was distorted by hand to produce the vowels while the other hand diverted air to different chambers that could make consonant-like noises. While the output was probably no better than the kind of Internet video in which a "talking dog" saying "sausages" as its owner manipulates its jaw, von Kempelen's speaking engine could produce recognizable words.

Improved mechanical versions were worked on occasionally through the nineteenth and early twentieth centuries. The most notable success was the vocoder, developed by Bell Labs in the late 1920s and early 1930s, which used manually controlled oscillators and filters to modify a noise source to produce speech. The idea was that speech could be encoded, sent as a set of unintelligible control signals, and reformed at a remote location by a vocoder. This could be used either to encrypt a message or to compress it to use a tiny fraction of the bandwidth that normal speech transmission required, but the overly complex system never caught on outside niche applications.

It was only with the advent of the electronic computer that the possibility of going far emerged further. The idea of making computers produce sounds began very early in the history of information technology. The great computing pioneer Alan Turing wrote the first program to generate music on the electronic computer at the University of Manchester. The machine was rigged up with a speaker to produce an alarm signal when something went wrong. Turing realized he could program this to issue tiny clicks. If these were put out thousands of times a second, it would produce a musical note that varied with the frequency of the clicks. Despite frequently being attributed to Bell Labs in 1957, the first computer music was produced using Turing's manual on programming the Manchester ACE computer (he himself only used the ability to produce different tones to give feedback) as early as 1950.

At the time that *2001* was filmed, though a lot of work had been done on understanding the components of speech and how they are produced, little progress had been made on artificial-speech generation. But from the 1970s onward, as computing power got far cheaper, providing a wider market for the potential applications of a speaking computer, progress was swift, through to the 1970s originated device used by Stephen Hawking and on to today's sophisticated capabilities. There's still a way to go before a computer can be as smoothly human as Hal (who was, after all, voiced by an actor), but we are well down the road.

While we may have impressive synthetic speech, getting a computer to accurately transcribe what a human says and to act appropriately in response to those words has proved a far bigger obstacle. It might seem that this too is a problem that we've cracked. After all, not only can my smartphone's navigation app speak directions to me in an easy-to-understand fashion, it also has Siri, a computerized personal assistant, who can follow my voice instructions. And that's just in a phone, without Hal's massive structure and presumably equally massive storage and processing capabilities.

Examine Siri's capabilities a little closer, though, and it becomes clear that she is a limited conversationalist. Admittedly, the programmers have anticipated some of the obvious remarks and come up with entertaining replies. When I tried, "Open the pod bay door, Hal," Siri responded, reflecting the scene in the movie, where the character Dave Bowman is stranded by Hal without a helmet for his space suit, "Without your space helmet, Brian, you're going to find that rather . . . breathtaking." When asked to sing "Daisy, Daisy," as the dying Hal does toward the end of the movie, Siri came back with: "You wouldn't like it."

Over and above a few novelties and handling the kind of request you might expect to make of an electronic PA—booking appointments, looking something up online, planning a route, or playing music—it rapidly becomes clear that Siri is not capable of a real conversation, falling down on both of the key challenges of parsing and of understanding speech. Siri's voice recognition is surprisingly effective, but there are times when it can struggle. Nonstandard accents can throw it easily—there isn't a speech-recognition system yet that doesn't fail with some Glasgow or Downeast Maine accents. And the way we speak as a matter of course involves slurring, running words together in a way that we never notice, but which a machine is forced to encounter and deal with.

This doesn't mean that it is not possible for a machine to understand speech. **I have just dictated this sentence into my Mac using the built-in software, the factors as you can see it can slip up.** "The factors as"? What I actually said was: "But the fact is, as you can see, it can slip up." Short connecting words like "but" are often truncated to hardly anything, while "fact is" and "factors" sound very similar with an English accent. Computer dictation software has become good enough that the best examples can cope with perhaps a 99 percent success rate. But that still leaves the computer more susceptible to error than a human. And to get the best out of dictation usually involves training the software to recognize your individual vocal quirks.

One of the problems that a computer faces is that it doesn't understand context. When we listen to a person speaking, we are constantly putting the words into the context of the wider conversation. Some words sound identical and have to be interpreted this way. This is particularly important if attempting automated translation between languages. If I am just using a computer to dictate text, it's easy enough for me to keep an eye out for such errors, but in an automatic translator I wouldn't know if a mistake was being made. Yet such a machine translator is something that we often see in science fiction, whether it's the universal telepathic translation provided by the TARDIS in *Doctor Who,* or the more realistic simultaneous computerized translation in *Star Trek.* Translation should, surely, be a major role for computers with language skills.

Hal never acts as a translator, but I am sure that he could. Google Translate on a phone can approximate to this. I can speak a simple phrase like: "Where is the nearest supermarket?" into the phone and have it display on the screen or speak aloud: *"Où est le supermarché le plus proche?"* for me and I can be reasonably confident that I'm getting what I ask for. The phrase might not be perfectly idiomatic, but I will be understood. However, should the conversation get significantly more complex, it would become dangerous to trust Google's software. We certainly aren't at the stage yet where the United Nations can rely on a machine translation while negotiating treaties.

Although Alexander Graham Bell and others played with mechanical ways of breaking down speech (and, of course, Roger Bacon's brass head and its mythical predecessors were supposed to be able to understand as well as speak themselves), it took computers to make this a reality. Neatly, considering Bell's enthusiasm, it was at Bell Labs in 1952 that a group made the first workable speech recognizer. The good news was that with training, this device managed 97 percent accuracy. The bad news is that it could only handle digits—well behind the kind of speech recognition program we tend to curse in modern automated telephone systems.

It is interesting that when speech-recognition pioneer Ray Kurzweil was writing about Hal's capabilities back in the late 1990s, he expected that we would be using speech to dictate to personal computer applications as the norm before 2001. In reality, such systems are still not used on most computers today. The shift Kurzweil expected has been much slower than he anticipated and may never come. Although the speech recognition technology built into my iMac is quite good, I very rarely use it. Many aren't even aware that it exists.

This is because the pioneers of speech recognition were so focused on the intellectual challenges of decoding speech that they forgot that such a system also needs to do something useful. Speech is great to ask a question like, "What's in my schedule for tomorrow?" but it is much easier to enter text into a computer with a keyboard unless you have a disability. For most of us, computer dictation is an uncomfortable process because, unlike a human secretary taking dictation, the software can't allow for the way we speak in fits and starts, changing our minds as we go. Most of us don't visualize sentences whole and neat, ready to be dictated.

Things would change significantly if software could not only convert speech to text, but interpret what it means—but that further phase is harder still. Hal doesn't just come up with a text version of the speech that he hears but parses it, understands what is being said, and formulates a response, or makes something happen. The vagaries of voice-command systems in cars frequently form a running gag in TV comedies, because speech is so easy to misunderstand. It is one thing to say something clearly into a phone's microphone, another to say a command over road noise and while distracted. In usual conversational circumstances, interpreting even the simplest requirements can be a minefield.

Unlike dictation, a computer that truly understood speech would genuinely be useful. The time when ex-Pepsi CEO John Sculley ran Apple Computer in the 1980s has often been criticized by those who see true creativity only returning to the company with Steve Jobs. But it was under Sculley's watch in 1987 that Apple came up with what is arguably its most visionary concept in a video demonstrating a fictional product called Knowledge Navigator. The product featured an electronic personal assistant, but unlike Siri, this one could deal with complicated requests like "Let me see the lecture notes from last semester," or, on being requested to provide journal articles, was able to point out that a friend of the owner had published a relevant article. Some aspects of the video, like the tedious recitation of calendar entries already seem dated, but much of Knowledge Navigator shows what would truly be a remarkable work aid if a device could reliably parse conversational speech and act accordingly.

A computer like Hal, though, has to manage far more than Knowledge Navigator's limited work-based interactions. To manage general conversation, Hal has to respond as much as possible as a human being would. Here, context is even more important than getting the words right. A phrase that has fascinated computer speech experts for decades is: "Fruit flies like an apple." It's easy enough to dictate this—I spoke it into my Mac just now. But what does it mean? Taken out of context it is impossible to say. It could be describing the eating habits of a particular type of fly, or the aerodynamics of fruit.

While that specific sentence is unlikely to come up in any imaginable conversation, there is no doubt that the whole process of understanding what people say is fraught with difficulty. Alan Turing envisaged a test to see if a computer could be considered intelligent. The device would be isolated in a room and a person would interrogate it from a second room, trying to decide if the "person" on the other end of the line is human or machine. (Turing's original statement of his idea was more complex, but this is the important part of it.) For decades since, computer scientists have been trying to beat this so-called Turing Test, and you will regularly see news items saying that it has been achieved. They are being generous with the truth. The Turing Test hasn't been beaten and is still probably a decade or two away from successful completion.

While Hal could indubitably win the Turing Test (I'm not so sure about the taciturn astronaut, Dave), the actual conditions under which competitions based on the test take place are far too trivial to demonstrate any degree of certainty. These events couldn't possibly test for the kind of human-like capacity that Turing had in mind. All that is usually required to "win" the test is that in a series of five-minute sessions the computer is mistaken for a human at least 30 percent of the time. The software written to try to win these competitions is *not* designed to be good at human conversation, but to be good at passing the test, often by being evasive.

The grandmother of all such programs is one known as ELIZA. This was written back in the 1960s, two years before *2001* was released, and is a tiny piece of code that nonetheless manages a reasonably sensible, if often frustrating conversation. Here the main evasion is provided by the pretense that ELIZA is a psychotherapist, and so she resorts to repeating a lot of what you say back to you, without giving much of an answer herself. Bearing in mind that ELIZA consists of fewer than 400 lines of code, this is still quite remarkable. You can try out a modern implementation of ELIZA at my website www.universeinsideyou.com/experiment10.html.

Such programs have moved on since. (It's hardly surprising, given that at the time of writing, ELIZA is approaching her fiftieth birthday.) In 2014, much was made of an apparent win of the Turing Test by a program called Eugene Goostman, which simulated a thirteen-year-old Ukrainian boy whose lack of English as a first language was one of the techniques used to evade detection. I couldn't test the Goostman chatbot (as these programs are called) myself, as it has been strangely unavailable since it was supposed to have won, but here is a short conversation I had with one of its leading competitors, Cleverbot:

Brian: *Hello, how are you?*
Cleverbot: *I'm fine, how about you?*
Brian: *Not bad but it's too hot.*

Cleverbot: *Why don't you like it hot?*
Brian: *Because it's uncomfortable.*
Cleverbot: *That's lame!*
Brian: *Why?*
Cleverbot: *How old am I?*
Brian: *How should I know?*
Cleverbot: *Because you are my phone!*
Brian: *What do you think of the situation in Syria?*
Cleverbot: *What situation?*
Brian: *Don't you watch the news?*
Cleverbot: *I don't have a TV.*

There's some classic evasion at the end there, denying a source of information that would be very unlikely in a real person and shutting down potential conversational hot spots that would show that Cleverbot was definitely not a person. It's unusual for a human to have no TV and to have no knowledge that anything was happening in Syria where, at the time of writing, a war was a major news item. But that center section of the conversation provides us with a classic chatbot giveaway that immediately proves this is no human. The software responds to "Why?" with a non sequitur, and gets totally confused when asked how I would know its age. What I find difficult to believe is how many judges do seem to be fooled by this kind of clever programming that is easy to draw into a conversational black hole without any tricks. I have never yet had an online conversation with a chatbot lasting five minutes where it hadn't given itself away a number of times in this fashion.

Of course, Hal has far better conversational skills than a chatbot. He is not just trying to fool us to pass a test, but is truly capable of a real conversation. We get the impression from the way that he speaks that he has intelligence—artificial intelligence

(AI), but something with a strong parallel to the human capabilities that we would typically label as intelligence. And that is the biggest challenge of all those faced in matching the capabilities of Hal. Ever since the 1980s there have been attempts to produce something that has become called AI, with two broad thrusts. One was expert systems and the other neural networks. (There were, and remain, a wide range of other possibilities, but these two probably best typify the differing attempts to give computers a form of intelligence.)

Expert systems come at intelligence with a leech-like concept of draining a human of their expertise and accessing it. (As you might expect, the companies marketing expert systems presented them in a better light.) Through a process known as knowledge elicitation, also known as asking a lot of questions in a structured way, the idea was to extract the essentials of expertise in a particular subject area from one or more experts and put the information into a sophisticated database which could then replace a fragile human with an electronic substitute that could work 24/7 and serve as many clients at a time as the computer could make connections. Looking back, it's hard to see how anyone ever believed this approach could be effective as a way of reproducing real expertise.

The problems stacked up quickly. Firstly, there was the collaboration of the experts. It's one thing for someone to come up with a simple, confined, and well-known piece of knowledge, such as "how to boil an egg"—the kind of thing that was probably used in the initial academic testing of this so-called knowledge elicitation process. But take, say, a medical doctor or a rocket scientist and attempt the same thing and you quickly realize that you are going to have problems. Even the most cooperative expert can easily get bogged down when trying to explain what they do, as the chances are they don't understand at a conscious level all of what their expertise entails. But realistically, why would an expert want to be truly cooperative even if they could be? The builders of expert systems might argue that the system would free up the expert from routine, boring work, leaving him or her to concentrate on the interesting bits. But experts are just as likely to be suspicious of machines that it is claimed can do their jobs as are manual workers. It's unlikely that the "expertise" that was elicited was top quality.

Then there is the sheer complexity of such expertise. It's not like the egg-boiling example, where a simple set of instructions and rules can pretty well ensure a perfect outcome. The nature of the knowledge involved is far more than a set of information and a framework. Knowledge involves knowing what to do with the information provided—which entails linking it to all sorts of other information and expertise. Arguably it involves deploying creativity. And this complex network of information, inspiration, and relationships proved pretty well impossible to extract and structure. There was far too much to cope with, incorporating far too many interrelations and links.

Expert systems didn't die out entirely, but ended up restricted to more sophisticated versions of the "how to boil an egg" expertise. You are most likely to come across such systems in the diagnostic scripts used by call centers and in some computer apps, which attempt to work out why a piece of technology isn't working. But the experts of this world were able to rest happy in their beds. A real-life HAL 9000 based on expert system technology was never likely to replace them because their expertise could not be extracted and structured into a database.

Expert systems were an attempt to apply logic and hierarchical organization structure to knowledge, but we know that the human brain is a very different kind of system from a traditional database—and the opposing style of AI, neural networks, which have had more lasting value, attempt instead to reproduce the kind of mechanism used by the brain in computer form. Instead of devising a series of rules, the designer of the network mimics the way that neurons in the brain function. Between the requirement that is input to the neural network and the result that is output are a series of virtual neurons. These are a bit like alternative routes through a special multiply connected maze, in which the computer can take all routes at once. The different connections have "weightings" that make them contribute more or less to the decision outcome.

These weightings will initially either be equal or will be set to what seems to the designer like a reasonable balance—but the thing that makes a neural network impressive and reminiscent of a living creature is that it can learn from its "mistakes." If the artificial intelligence program makes an incorrect decision in the learning process, it is informed of the best output and the weightings are automatically changed to reflect this. Over time, the network gets tuned to the levels that are required to trigger a particular output, so that it produces the "correct answer" more of the time. In effect, a neural network evolves its output in response to feedback.

As yet, no neural network has come close to the realities of the human brain. It's not surprising. The brain contains around 100 billion neurons, which at any one time will have around 1,000 trillion connections linking them. By comparison a typical neural network may only have hundreds or thousands of the "nodes" that correspond to neurons. And those nodes will typically only be connected to a handful of others, where neurons in the brain can have thousands of connections. The scale is totally different—and yet even with this very limited "brain," neural networks can work very well on small, focused areas of apparent intelligence, gaining a form of expertise as they pass through the learning process. Neural networks often prove effective where a traditional programming set of rules fails, particularly where pattern recognition (so important in vision) is involved.

It is clear now (though it wasn't in the 1960s) that a pure rule-based approach could never deliver a computer with Hal's abilities and that it would require some equivalent of a neural network approach, combined with many other structures and methods. It's interesting that when Hal is disabled toward the end of the movie he reverts to a child-like form, implying that he did have to go through a learning process and wasn't switched on fully formed. Only one HAL 9000 would have to go through that learning process. It should then be possible to duplicate the structures for others in the series—though the wording of the script for Hal's decaying mind suggests that the writers hadn't spotted this.

There is much debate about whether computers can be truly intelligent—or even conscious—or whether they can simply provide better and better simulations of intelligence, like the chess-playing computer simulates understanding the game of chess, without achieving this in reality. This strays somewhat from science into philosophy, but is still important for those working in the field to consider. There is no clear answer. Some argue that Hal certainly wasn't intelligent, at least by sensible definitions. He demonstrates actions that could only be described as lacking intelligence. At one point, for example, he says, "No 9000 computer has ever made a mistake." Yet almost by definition, in order to be intelligent, an entity, whether flesh and blood or machine, has to be able to make mistakes to learn from them.

There is a real problem with Hal's claimed perfection. Making mistakes is a fundamental part of the learning process, something that would be essential to enable a computer like Hal to develop intelligence. So there should be mistakes along the way. Not in a traditional computer arena like mathematical calculation, admittedly. That's an area where humans are always making mistakes and computers never have a need to fail. Even a dumb calculator can make this claim. But in the whole area of decision making and communication with others, mistakes are inevitable because the intelligent device has to act based on incomplete information. Hal could only make the claim to be incapable of error if he had access to perfect information, and to make that claim in the real world implies an absence of intelligence.

For that matter, Hal's homicidal actions are not only irrational, but arguably lacking in intelligence. In the movie, Hal suffers a conflict because he has orders to keep the real reason for the mission secret. In his attempt to resolve this stress, he kills off members of the crew. But this is hardly the most effective way of achieving his goals. Hal seems to lack significant parts of what we would consider is involved in being intelligent—and parts that furthermore should be relatively easy to program.

Science fiction hasn't just put a computer in charge of a spaceship as in *2001*. In some cases we see societies run by computers, either as the ultimate electronic dictator, or providing Delphi-style government (more on Delphi in a moment). So, in James Blish's Cities in Flight series, where cities abandon the wasted Earth using a technology that allows the whole city to be launched into space, there are a series of computers called the "City Fathers." The description of these machines is quaintly hilarious. The City Fathers are antiquated in conception, large metal boxes that have to maneuver on tracks and physically connect with each other to exchange information. No networks for them. But they manage to provide a kind of benign dictatorship.

Although each city is run by a human mayor, it is down to

the City Fathers to decide if he or she is doing a good enough job. And if that's not the case, they simply execute or exile the individual concerned. Blish was trying to provide a mix of human direction, but with the steely and totally honest computer as a balance to ensure that nothing goes astray. By contrast, we see the Delphi approach, typified by the John Brunner novel *The Shockwave Rider*, a novel with a more modern conception of information technology. Brunner uses the computer's ability to combine ultrafast communication and data processing to transform the nature of democracy, making government something that resembles a cross between online polls and a casino, with its collation of human responses and betting odds.

The book's title is a reference to Alvin Toffler's stodgy 1970 work of futurology, *Future Shock*. While Toffler has proved pretty well universally wrong when it comes to the real world, his ideas were much followed at the time and proved a great inspiration for fiction writers. In this case the concept is built on the Delphi method, originated by the RAND Corporation in the 1950s. In the original Delphi process a series of individuals were given a decision to make, or questions to answer. After a first pass at answering, the results were fed back to the whole panel, who then had the opportunity to revise their answers. There seemed some evidence that the result of this procedure was to home in on a better solution than would have been achieved by just taking the original opinions of the decision makers.

In Brunner's book, inspired by Toffler, this Delphi approach becomes the means of government. Rather than have decisions made by a few hundred elected representatives, the country's legislative body is provided by the population as a whole. Their immediate thoughts are fed back to them in a form resembling a betting process, through networked systems. With repeated Delphi-style consultation, the best decision is (supposedly) made.

Although it is humans that provide the intelligence at the individual level, it is only because of the computer network, tying together their decisions and feeding them back, that this approach is viable. By bringing the computer into the process, the ultimate in democracy is achieved—provided that the system is secure and properly programmed. Whether or not such democracy is what we want is, of course, a different thing. The distinction between democracy and mob rule is an interesting one that modern electronic communications make us really think about.

There is no real sign of computers actively becoming involved in running our world, but there is no doubt they have made a massive change to the way we live, a change that has crept in almost unnoticed over the years. Look and marvel at the way that the computer-based Internet furnishes us with information and entertainment, shopping and socializing. Computers have, in their way become our interface to the world. You only have to think about at the way you might use your pocket computer (as a smartphone truly is) to find your way across a city, locate a coffee shop and pay your tab, communicate with friends, and entertain yourself during a wait. Computers certainly have a huge influence on how we go about our daily lives.

A simple example is the production process for this book. I have never met my editor, who is based 3,500 miles away. The book itself was written on a computer in my home, backed up on cloud servers that could be anywhere in the world. The different versions of the text and e-mails discussing them have flown backward and forward across those huge distances time and again. And as more and more of us make use of e-books, it is entirely possible that the copy you are now reading may have gone from my computer via various electronic forms to the tablet or e-reader you are consuming it on without ever being printed on paper.

Science fiction has brought us many visions of what may come to pass, with and without computers at its heart. That is part of the appeal of the genre—the breadth of possibilities is breathtaking. We are not limited to what we know, or even what we think is possible, but can explore "What if?" to our hearts' content. And that is something that is likely to continue as long as books, TV, movies, . . . and whatever technology can bring us to replace them continue to be made. So an inevitable last thought is to consider what will happen to science fiction in the future.

18.

NEVER-ENDING STORY

Although science fiction is a genre of innovation, once it hits on a good theme it is reluctant to let go. Broadly, these tropes divide into two areas. There are the handy MacGuffins that are unlikely to be real technology any time soon if ever—think of time machines, antigravity, faster-than-light drives, and instantaneous communication across any distance—and those where we can already see technological developments that parallel and in some cases far exceed the science fiction dream. In *Ten Billion Tomorrows* we have mostly stuck to these second cases, but we have to bear in mind that there will be new technologies of the imagination that come along. Science fiction may still rely on hoary old classics, but there is always new material coming in.

It's easy to be cynical about sci-fi technology. It's tempting to ask, "Where are the flying cars and the ray guns?" But the fact is that if we compare our real technology with that experienced by the early writers of science fiction—Victorians-era writers like H. G. Wells and Jules Verne, or with the first sci-fi moviemakers—we are living in a world that would indeed be science fiction to them, with technology that has transformed how we live. In this book we have discovered just how far we have come in so short a time, as the boundaries between science fiction and science fact fell away, even if the futures that the writers conceived rarely matched what has actually happened.

Part of the problem in looking for any such match is that it returns us to the misunderstanding mentioned in the opening chapter. Science fiction is rarely an attempt to predict what will become reality. The 2004 remake of the 1978 TV show *Battlestar Galactica* is an interesting reminder of how little interest prediction can have for the writers. As we have already seen, the original show was little more than a cynical attempt to make money out of the Star Wars phenomenon, but the remake was much more subtle, and if you look beyond the trappings of space opera—the huge spaceships and fiery space battles—what plays out is a much more sophisticated story than anything to emerge from either the Star Trek or Star Wars universes.

Consciously or not, this was reflected in what has to be the strangest decision in the setting of any future-based science fiction show. Once the trappings of a space empire—those huge ships and the ability to make faster-than-light jumps—are set up, the technology is solidly from the twentieth century. There are no phasers or blasters, just familiar, everyday guns firing bullets. The most terrifying weapons are nuclear warheads, not photon torpedos.

Communication technology is almost laughably dated—most of the phones are still connected to base units by cords. Books are books. Bits of paper are bits of paper. Okay, the designers couldn't resist making something a little different—so documents have an impractical shape with corners cut off, and playing cards are multisided instead of rectangular, but in essence, there is little that is used outside of the space gear that wouldn't be familiar today. The men even wear suits and ties.

It is possible this was all done to save money, or because the writers had limited imagination. But equally it could be that what is happening here is that we are being given the chance to ignore the technology and see past it to the really interesting part of the rebooted *Battlestar Galactica,* which is the interplay between the humans and the human-like Cylons, who started as a little more than a race of robots, but who have taken on a form with flesh that is almost indistinguishable from human beings. Without *Doctor Who*'s sonic screwdriver or *Star Trek*'s phasers and tricorders, it is easier to empathize with the humans in *Battlestar Galactica.* This is still a story line based on "What if?"—but imagined technology is not getting in the way.

There is a complex dance going on all the time between science fiction and technological developments in the real world. Science fiction writers need to make the practicalities of their story line work. When it comes to a restriction imposed by science, or current technology, they can do three things. The first possibility is to live with the restriction—to make it a fundamental part of the story.

We can see this in the different approaches taken to stories about interstellar travel. If a science fiction author decides to honor the limitation of not being able to travel faster than light, then they need to find alternative ways to enable human travelers to survive journeys covering vast distances at sublight speeds. So one approach might be to build vast generation starships in which hundreds of people will live and die over a journey time that might be hundreds or thousands of years. This approach cropped up in several classics, including Brian Aldiss's book *Hothouse,* in which the space travelers did not even realize that they were on board a ship.

Another example is to consider that if a ship can be pushed up to near light speed, the time elapsed on board will slow down due to special relativity, so the journey that may take hundreds of years will pass in a few months for the passengers. But this then brings in another interesting and challenging story line in the way that those on board have to cope with returning home to find all their friends and relations dead—a scenario expertly explored in Joe Haldeman's *The Forever War*.

Yet science fiction *is* indeed fiction. So should the author decide that it isn't necessary to stick to known technology, a second possibility is to extrapolate to one that might make the envisaged outcomes possible. It is this approach where science fiction does produce its best hits—and its worst mismatches compared with developments in reality. A good example of a technology that science fiction has explored at some length before the real world has achieved anything is the space elevator.

The concept of a space elevator is beautifully described in Arthur C. Clarke's novel *The Fountains of Paradise*. The idea is simple but powerful. One of the biggest obstacles to opening up space to easy travel is getting out of the Earth's gravity well. It takes a huge amount of energy to propel something massive into space. So why not run an elevator between the ground and orbit? That way it would be easy to build large space stations, simply by sending up the parts to assemble them, piece by piece. And solar system exploration vessels, or even starships, could be assembled in space, never having to experience the full might of Earth's gravity.

Clarke did not originate the idea. As far back as 1895, Russian space science pioneer Konstantin Tsiolkovsky proposed building a tower up to a geostationary orbit, which would contain a lift to haul things up, perhaps inspired by the Eiffel Tower in Paris. In practice this would never be feasible. A geostationary orbit—the distance above the Earth where a satellite orbits at the same speed as the Earth's rotation, and so stays above the same point on the ground—would need to be over 35,750 kilometers (22,250 miles) high. Such a tower would be impossible to build—it would collapse under its own weight.

This proved to be the spark for the kind of elevator described in Clarke's 1979 novel, based on an idea that had been played with by a number of scientists and engineers from the late 1950s onward. Instead of constructing upward from the ground, a space elevator is more like dropping a cable from a satellite and hauling objects up the cable. Even this has problems of scale. If the elevator were based on a wire rope (steel cable) 28 millimeters (or just over an inch) across, it would be capable of supporting a weight of around 50 tons. But the cable itself would weigh around 115,000 tons.

From a science fiction viewpoint, all we are faced with here is an engineering problem. There is nothing physically impossible about having a cable strong and light enough to actually work on a space elevator. We already have materials like Kevlar, which could just about make a space elevator function on the Moon, and in principle, carbon nanotubes could be made strong enough to operate an elevator from Earth. So the elevator was an excellent example of this second kind of science fiction approach, one that Clarke's writing often typifies.

It is entirely possible that space elevators will become useful, practical technology at some point, and there is currently a significant amount of work going into developing potential cable materials and solving the other technical problems a space elevator presents, such as how to get power to the motor that would crawl up the cable pulling the payload. (Unlike a traditional elevator, the cable itself can't sensibly move, so the elevator car or whatever makes the journey has to climb up a static cable.)

Equally, the idea of space elevators could become the equivalent of a *Star Trek* communicator, or the Pan Am space shuttles from *2001: A Space Odyssey*. All attempts to extrapolate into the future that seemed reasonable at the time, but that didn't turn out right. We could discover a whole new way of getting materials into space, or of dealing with the difficulties of the Earth's gravity well that would make a space elevator seem far too risky and slow. And however good our science fiction authors are at futurology, this is bound to happen—because we see the future through a veil of chaos.

A "veil of chaos" sounds like the weapon of a villain in a fantasy comic, but simply reflects the reality of trying to predict the future in a very complex system, such as the worldwide development of technology. The mathematical concept of chaos was devised when Edward Norton Lorenz, a meteorologist, was using an early computer-based weather model. He wanted to restart it partway through, so he entered the parameters from a printout, but was baffled when the model produced a totally different forecast to his previous run. It turned out the computer worked to six decimal places, but only printed out values to four places. That tiny difference in the starting conditions had produced a huge difference in outcome—which came to typify a chaotic system.

In theory, if you had every bit of information about a system in perfect detail you could predict the outcome, but in practice the systems we live with—the weather and the stock market are great examples—are so complex, and can have such different outcomes as a result of very small differences in starting conditions that prediction becomes impossible. This has such a powerful impact that even though you will see companies issuing ten-day weather forecasts, by the time we are looking ten days ahead, it is more accurate to make a prediction simply by describing what the weather is likely to be like at a particular location on that date than it is to attempt to use weather models to produce a true forecast.

If the weather system is so difficult to predict accurately, imagine how much more difficult the whole complex of human thought and interaction, of invention and scientific discovery is to predict. This is a shame really, as I based my early career on the inspiration of a fictional capability of doing just this. In Isaac Asimov's classic Foundation series, a mathematical genius called Hari Seldon, and the Foundation members who follow him, make long-term predictions of what is going to happen during the gradual disintegration of the galactic empire. They do this by using a complex mathematical system known as psychohistory, which relies on the observation that the behavior of crowds is much easier to predict than that of individuals.

When I finished my physics degree and was looking for where to go next, my eye was caught by a discipline called operational research (OR, known as operations research in the United States). It applied mathematical techniques, often statistical, to business problems, including forecasting. This seemed a way to make science fiction come true, and I went on to take a masters in OR before working for a number of years in the field at British Airways. OR is a very powerful discipline—but it can't come close to psychohistory, because the behavior of people is chaotic and impossible to predict, especially when it comes down to as complex a field as new inventions and scientific discoveries.

In Asimov's Foundation series books, an individual called the Mule breaks the system, as he is a mutant whose behavior is impossible for psychohistory to predict. What Asimov either forgot (or, more likely, chose to ignore—he was, after all, a biologist) is that we are all mutants, all different, none of us programmed to behave in a particular way. We are never going to be able to use mathematics as an accurate mirror to the future, whether attempting serious futurology or in the (arguably more valuable) field of science fiction.

Finally, when faced with devising a future technology, if plausible projections are not possible, the writer can bite the bullet and dismiss experience entirely, apply a patch to reality, and make pretty well anything happen. From the point of view of the book or movie, the important thing is now not whether or not what is described is possible, but, once the leap has been made, if the world as described is logical and consistent. Making this kind of leap is how we get from science fiction to fantasy. And even there, in the best fantasies, that consistency is applied. But more of interest to us is where the leap is merely a new scientific discovery or workaround that allows us to change the rules—but then sticks to them without the magical flexibility of fantasy.

It is here where we get those science fiction leaps of faith that we have already met at the start of the chapter, like time travel, faster-than-light travel, instantaneous communication, and antigravity. They will probably never come true, even in the very different ways that take us by surprise that we have met in this book, but they are useful shortcuts to make the main purpose of the story effective.

If you take a science fiction story written in recent years, there will be many facets of it that would be familiar to a reader from the 1940s. Take the excellent Culture series of novels by Iain M. Banks. In books like *Consider Phlebas* and *The Player of Games,* Banks constructs a whole far-future structure of planets and civilizations. The way he does this—the sophistication, if you like, of the story line—is far beyond the equivalent in a space opera of the 1940s. But a lot of the elements Banks used to construct his future world would be distinctly familiar. We have intelligent devices, large spaceships using faster than light drives, characters who can change sex, a warrior race with values we would consider medieval, and much more.

A twenty-first century writer has the advantage of knowing today's science and technology. They are unlikely to make the same errors about computers or phones that would be typical thirty or forty years ago. But they are still just as liable to diverge from the reality that will come in the next thirty or forty years. There are already scientific developments just waiting for writers to pick up on. Not enough, for instance, has been made of the possibilities of quantum entanglement (see page 167) or to think through the implications of room-temperature superconductors, which could revolutionize electrical technology if ever made possible.

Ten Billion Tomorrows happens to have been published within a few months of one of the best-known dates in science fiction "future history"—October 21, 2015. This is when Marty McFly arrives in the future thanks to Doc Brown's time-traveling DeLorean, set a comfortable thirty years ahead of *Back to the Future*'s 1985 release. The first movie only briefly visits 2015, but the sequel is primarily focused on the future, where we are given an object lesson in the pros and cons of science fiction as future gazing.

The most famous future technology of the second and third movies is probably the hoverboard, a floating skateboard. The version Marty uses is unlikely to become a reality, despite a real hoverboard appearing on the Kickstarter Web site in October 2014. The good news is that this "Hendo" board, designed by California start-up Arx Pax does work—but by using magnetic levitation (see page 146), which means it can only function when above a surface made of nonferrous metal, like copper or aluminum. So until we get metal sidewalks, we won't see hoverboarding kids on the streets.

When I saw the movies, something very different stood out for me—the information technology. (This could be because my only trick on a skateboard is to fall off it.) Some of it is hopelessly dated. In Marty's future they are still using laser disks (remember them?) and communicate with faxes. But the portrayal of the casual integration of computing and communication technology into everyday life was excellent. Marty is asked to sign a petition on a tablet computer (less stylish than the real thing, but still a good approximation to an iPad). In another scene, the future family is seated around the dinner table, but spend their time using portable technology to communicate with friends that aren't there. (The devices are glasses rather than smartphones, but the effect is the same.)

We can enjoy the mistakes in the way the *Back to the Future* movies portray 2015, and be impressed where they get it nearly right—but most interesting is that throwaway reflection on the impact of technology on a simple act like sitting down to dinner. *Good* science fiction will continue to help us think about the implications of our science and technology and how it could change human experience in the future. And as long as it continues to do so, we can be thankful for the mental challenges from science fiction's ten billion tomorrows.

NOTES

I.

THROUGH A GLASS, DARKLY

PAGE 3—Arthur C. Clarke's prediction of geostationary communication satellites was in Arthur C. Clarke, "Extra Terrestrial Relays," *Wireless World* (October 1945): 305–8.

PAGE 3—The early predictions of nuclear war and atomic bombs came from H. G. Wells, *The World Set Free* (London: Corgi, 1976).

PAGE 4—Ernest Rutherford's dismissal of atomic power was quoted in an article in the *New York Herald Tribune*, September 12, 1933.

PAGE 5—The identification of Verne's error in comparing his work with that of H. G. Wells is pointed out by Adam Roberts in Vassili Christodoulou's interview with Roberts, "In Praise of Sci-Fi," IAI News, accessed September 5, 2014, iainews.iai.tv/articles/in-defence-of-sci-fi-auid-385.

PAGE 9—Hugo Gernsback's definition of scientifiction is quoted in John Clute and Peter Nicholls, *The Encyclopedia of Science Fiction* (London: Orbit, 1999), p. 311.

2.

BLUE PILL OR RED PILL?

PAGE 15—The story in which "cyberspace" is first used by Gibson was William Gibson, "Burning Chrome," *Omni* (July 1982).

PAGE 18—Information on John Carmack's inspiration by the *Star Trek* holodeck is from David Kushner, *Masters of Doom* (London: Piatkus, 2003), p. 25.

PAGE 24—The experiment relaying images from a cat's retina onto a screen is described in G. B. Stanley, F. F. Li, and Y. Dan, "Reconstruction of Natural Scenes from Ensemble Responses in the Lateral Geniculate Nucleus," *Journal of Neuroscience* 19, no. 1 (1999): 8036–42.

PAGE 25—The four books in James Blish's *Cities in Flight* sequence are: *They Shall Have Stars* (London: Arrow, 1974), *A Life for the Stars* (London: Arrow, 1974), *Earthman, Come Home* (London: Arrow, 1974), and *A Clash of Cymbals* (London: Arrow, 1974).

PAGE 29—The 1956 research showing no value from sleep learning is described in Charles Simon and William Emmons, "The EEG, Consciousness and Sleep," *Science* 124 (1956): 1066–69.

PAGE 29—Research showing that memory consolidation could be added during sleep with an odor is described in Susanne Diekelmann, Christian Büchel, Jan Born, and Björn Rasch, "Labile or stable: opposing consequences for memory when reactivated during waking and sleep," *Nature Neuroscience* 14 (2011): 381–86.

PAGE 29—Research linking hearing music during sleep to better recall is Thomas Schriener and Björn Rasch, "Boosting Vocabulary Learning by Verbal Cueing During Sleep," *Cerebral Cortex* (2014), doi: 10.1093/cercor/bhu139, accessed September 29, 2014.

PAGE 30—Research showing improvement of memory consolidation during sleep with transcranial stimulation is Lisa Marshall, Matthias Mölle, Manfred Hallschmid, and Jan Born, "Transcranial Direct Current Stimulation During Sleep Improves Declarative Memory," *Journal of Neuroscience* 24, no. 44 (November 3, 2004): 9985–92, doi: 10.1523/JNEUROSCI.2725-04.2004, accessed September 29, 2014

PAGE 32—The figure of more than $2 trillion as the cost of another Carrington event is quoted in a NASA Science News piece, accessed July 30, 2014, science.nasa.gov/science-news/science-at-nasa/2014/02may_superstorm/.

PAGE 34—The use of high-altitude nuclear explosions to generate electromagnetic pulses (EMPs) is discussed in Edward Savage, James Gilbert, and William Radasky, *The Early-Time (E1) High-Altitude Electromagnetic Pulse (HEMP) and Its Impact on the U.S. Power Grid* (January 2010), available at web.ornl.gov/sci/ees/etsd/pes/pubs/ferc_Meta-R-320.pdf.

PAGE 35—Larry Niven first used the term "wirehead" in the novella *The Defenseless Dead,* published in the collection Roger Elwood (ed.), *Playgrounds of the Mind* (New York: Fawcett 1973).

PAGE 35—The idea of artificial intelligences tending to become wireheads is raised in Nick Bostrom, *Superintelligence* (Oxford, UK: Oxford University Press, 2014), p. 122.

3.

BUY ME

PAGE 37—Quotes on science fiction advertising and marketing methods are from Frederik Pohl and C. M. Kornbluth, *The Space Merchants* (London: Penguin Books, 1965), pp. 9–10.

PAGE 38—The more aggressive Mokie-Koke advertising is described in Frederik Pohl, *The Merchants' War* (London: Futura, 1986), pp. 47–49.

PAGE 40—Claims of James Vicary's success in the use of subliminal advertising were made in "'Persuaders' Get Deeply 'Hidden' Tool: Subliminal Projection," *Advertising Age,* September 16, 1957, 127.

PAGE 41—One of several studies showing a preference for a particular brand but an inability to prime for thirstiness using subliminal means is Johan C. Karremans, Wolfgang Strobe, and Jasper Claus, "Beyond Vicary's Fantasies: The impact of subliminal priming and brand choice," *Journal of Experimental Social Psychology* 42, no. 6 (November 2006): 792–98.

PAGE 43—Details of the Spanish poster campaign with different messages at adult and child's height are from David Kiefaber, "Child-Abuse Ad Uses Lenticular Printing to Send Kids a Secret Message That Adults Can't See," *Adweek,* May 6, 2013, accessed October 8, 2014, www.adweek.com/adfreak/child-abuse-ad-uses-lenticular-printing-send-kids-secret-message-adults-cant-see-149197.

4.

FEEL THE FORCE

PAGE 50—Details of the inclusion of a form of tractor beam in a posthumous Verne book are from Stephen R. Wilk, *How the Ray Gun Got Its Zap* (Oxford, MA: Oxford University Press, 2013), p. 220.

PAGE 50—The first use of "tractor beam" can be found in E. E. Smith, *Spacehounds of IPC* (Reading, UK: Fantasy Press, 1947), p. 117.

PAGE 51—The Australian water-based tractor beam is described in Jonathan Webb, "Physicists make 'tractor beam' in water," BBC News, accessed August 12, 2014, www.bbc.co.uk/news/science-environment-28739368.

PAGE 53—For a demonstration of a Crookes radiometer or light mill in action see www.universeinsideyou.com/experiment4.html, accessed August 6, 2014.

PAGE 53—Details of possible optical technologies for real, small-scale tractor beams are from Aristide Dogariu, Sergey Sukhov, and Juan José Sáenz, "Optically induced 'negative forces.' *Nature Photonics* 7, no. 1 (2013): 24–27.

PAGE 54—The tractor beam using a donut-shaped laser beam and glass spheres is described in Vladlen Shvedov, Arthur R. Davoyan, Cyril Hnatovsky, Nader Engheta, and Wieslaw Krolikowski, "A long-range polarization-controlled optical tractor beam," *Nature Photonics* (2014), doi: 10.1038/nphoton.2014.242.

PAGE 56—Details of the magnetic "tractor beam" for nonmagnetic objects is from Maria Dasi Espuig, "Scientists manipulate magnetically levitated objects," BBC News, accessed September 5, 2014, www.bbc.co.uk/news/science-environment-28904406.

PAGE 57—The article claiming that we are beginning to get holodeck technology was Nick Bilton, "The Holodeck Begins to Take Shape," *New York Times*, January 27, 2014, B5.

PAGE 60—Realistic full-color images in holograms can be seen on the Colour Holographic Web site, accessed September 24, 2014, www.colourholographic.com.

5.

ROSSUM'S CHILDREN

PAGE 65—*R.U.R.* is available in translation by the author in Karel Čapek, *R.U.R. & War with the Newts* (London: Orion, 2011).

PAGE 65—The early uses of the word "android" to mean a

mechanical human are referenced in the *Oxford English Dictionary*, accessed September 5, 2014, www.oed.com/view/Entry/7333.

PAGE 66—Information on Leonardo da Vinci's automata from Marina Wallace (ed.), *30-Second Leonardo da Vinci* (Lewes, UK: Ivy Press, 2014), p. 90.

PAGE 70—For more on the science fiction portrayal of the nanotechnology threat, see Michael Crichton, *Prey* (New York: Harper Collins, 2002).

PAGE 70—For more on nanotechnology, nanobots, and assemblers, see K. Eric Drexler, *Engines of Creation* (New York, Bantam, 1986).

PAGE 71—A transcript of Feynman's talk is available at Richard Feynman (1959), *Plenty of Room at the Bottom* [online] MSU Department of Physics and Astronomy. Available at www.pa.msu.edu/~yang/RFeynman_plentySpace.pdf, accessed February 12, 2015.

6.

DINOSAUR CONSTRUCTION

PAGE 79—The quote on grinding dinosaur bones is from Michael Crichton, *Jurassic Park* (London: Random House, 1993), p. 63.

PAGE 79—The discovery of T. rex hemoglobin is detailed in Mary H. Schweitzer et al, "Heme compounds in dinosaur trabecular bone," *Proceedings of the National Academy of Science of the Unites States of America* 94 (June 1997): 6291–96.

PAGE 79—The quote on finding blood-sucking insects in amber is from Michael Crichton, *Jurassic Park* (London: Random House, 1993), p. 95.

PAGE 80—Details of the experiment determining the absence of DNA in insects in copal are from "The final nail in the Jurassic Park coffin," University of Manchester News, September 12, 2013, accessed September 29, 2014, www.manchester.ac.uk/discover /news/article/?id=10630.

PAGE 80—Information on the half-life of DNA is from Matt Kaplan, "DNA has a 521-year half-life," *Nature,* October 10, 2012.

PAGE 85—Details on current state of mammoth research are summarized in Jeanne Timmons, "Could Ancient Giants Be Cloned," *Valley News,* January 7, 2013, accessed September 30, 2014, www.vnews.com/home/3694233-95/cloning-mammoth -mammoths-extinct.

7.

SUIT UP

PAGE 90—For more on the history of rocket belts, see Paul Brown, *The Rocketbelt Caper* (Amazon Kindle, 2009).

PAGE 90—Isaac Asimov's prediction of commuting by rocket belt is from "Life in 1990," *Science Digest,* August 1965, 63–70.

PAGE 94—Information on jumping stilts from Web sites including www.pro-jump.co.uk and getjumpingstilts.com, accessed October 1, 2014.

PAGE 95—Information on the SpringWalker is from the Web site www.springwalker.com, accessed October 1, 2014.

PAGE 96—Information on DARPA's Exoskeletons for Human Performance Augmentation program is from the DARPA Web site www.darpa.mil/dso/thrust/matdev/ehpa.htm, no longer available but accessed October 1, 2014, via web.archive.org/web /20070702155658/http://www.darpa.mil/dso/thrust/matdev /ehpa.htm.

PAGE 97—Details of the BLEEX exoskeleton are found in

H. Kazerooni, "The Berkeley Lower Extremity Exoskeleton Project," (paper presented at the Proceedings of the 9th International Symposium for Experimental Robotics, Singapore, June 2004).

PAGE 97—Information on the TALOS "Iron Man suit" request for proposals from www.army.mil/article/113332/_Iron_Man__style_suit_in_early_stages_of_development, accessed October 1, 2014.

PAGE 98—Information on DARPA's "Warrior Web" program from www.darpa.mil/Our_Work/BTO/Programs/Warrior_Web.aspx, accessed October 1, 2014.

PAGE 99—The artificial muscles are described in Von Howard Ebron et al, "Fuel-Powered Artificial Muscles," *Science* 311 (2006): 1580–83.

8.

RAY GUN READY

PAGE 104—The MythBusters' attempts to recreate Archimedes' concentrated sunlight weapon, and alternative scientific trials are discussed in Stephen R. Wilk, *How the Ray Gun Got Its Zap* (Oxford, UK: Oxford University Press, 2013), pp. 12–23.

PAGE 104—The derivation and earliest known usage of "ray" in English is from the *Oxford English Dictionary,* accessed September 27, 2014, www.oed.com/view/Entry/158718.

PAGE 105—Details of early fictional ray guns are from Stephen R. Wilk, *How the Ray Gun Got Its Zap* (Oxford, UK: Oxford University Press, 2013) pp. 198–207.

PAGE 106—Details of Nikola Tesla's life and work from W. Bernard Carlson, *Tesla: Inventor of the Electrical Age* (Princeton, NJ: Princeton University Press, 2013).

PAGE 107—Tesla's claim that his weapon could bring down a fleet of 10,000 airplanes was reported in "Tesla, at 78, bares new 'death beam,'" *New York Times,* July 11, 1934, 18.

PAGE 108—The suggestion that Gene Roddenberry changed the name of the laser to "phaser" as he didn't want complaints that lasers couldn't stun people is from Stephen R. Wilk, *How the Ray Gun Got Its Zap* (Oxford, UK: Oxford University Press, 2013), p. 236.

PAGE 111—Information on the development of the laser from Brian Clegg, *The Quantum Age* (London: Icon Books, 2014), pp. 132–50.

PAGE 115—The light saber headlines quoted are from Fox News, September 29, 2013; *The Guardian* newspaper, October 1, 2013; and *Time* magazine, October 1, 2013, respectively.

PAGE 115—Professor Lukin's "It's not an inapt analogy" remark was reported in Peter Reuell, "Seeing light in a new way," *Harvard Gazette,* September 27, 2013.

9.

TAKE ME TO YOUR LEADER

PAGE 117—The description of the Martian is from H. G. Wells, *The War of the Worlds* (London: Penguin Books, 1956), pp. 132–33.

PAGE 124—Stephen Hawking's warning about the danger of contacting aliens is described in Brian Clegg, *Final Frontier* (New York: St. Martin's Press, 2014), pp. 82–83.

PAGE 133—Quotes from the Drake equation are taken from Frank Drake (1961). *The Drake Equation | SETI Institute.* [online] Seti .org. Available at www.seti.org/drakeequation, accessed February 12, 2015.

10.
END OF THE WORLD, PART THREE

PAGE 138—Sue Guiney's reminiscence of air raid drills is from her blog, www.sueguineyblog.blogspot.com.

PAGE 139—John Wyndham, *The Chrysalids* (London: Penguin, 1970).

PAGE 139—Walter M. Miller Jr., *A Canticle for Leibowitz* (London: Corgi, 1975).

PAGE 141—Brian Clegg, *Xenostorm: Rising* (Golden: Reanimus Press, 2013).

11.
ATOM WORLD

PAGE 146—For more on superconductors and magnetic levitation see Brian Clegg, *The Quantum Age* (London: Icon Books, 2014), pp. 173–200.

PAGE 146—Details of floating frogs and antigravity are from Brian Clegg, *Gravity* (New York: St. Martin's Press, 2012), pp. 250–80.

PAGE 152—The comparison of energy density of different fuels is taken from Richard A. Muller, *Physics for Future Presidents* (New York: W. W. Norton, 2008), p. 69.

PAGE 153—For details of different types of spaceship propulsion see Brian Clegg, *Final Frontier* (New York: St. Martin's Press, 2014).

PAGE 154—Details of Project Orion from G. R. Schmidt, J. A. Bonometti, and P. J. Morton, "Nuclear Pulse Propulsion—Orion and Beyond," (paper presented at the 36th AIAA/ASME/SAE/ASEE Joint Propulsion Conference, Huntsville, AL, July 2000).

PAGE 157—Information on Harold White's Alcubierre warp drive design is from Harold White, "Warp Field Mechanics 101," NASA Archive, accessed September 9, 2014, ntrs.nasa.gov/archive/nasa/casi.ntrs.nasa.gov/20110015936_2011016932.pdf.

12.

BEAM ME UP

PAGE 160—Charles Fort's coining of the word "teleportation" was in Charles Fort, *Lo!* (New York: Cosimo Inc., 2006), p. 53.

PAGE 162—The first reference to the Mary Celeste as "Marie Celeste" is in "J. Habakuk Jephson's Statement," collected in Arthur Conan Doyle, *The Captain of the Polestar and Other Stories* (Auckland, New Zealand : The Floating Press, 2010), pp. 44–87.

PAGE 163—Details of the Lan Wright novel "Transmat" are from John Clute and Peter Nicholls, *The Encyclopedia of Science Fiction* (London: Orbit, 1999), p. 787

PAGE 168—For more on quantum teleportation see Brian Clegg, *The God Effect* (New York: St. Martin's Griffin, 2009), pp. 205–19.

13.

DESTINATION MOON

PAGE 170—For more on other aspects of space travel, see Brian Clegg, *Final Frontier* (New York: St. Martin's Press, 2014).

PAGE 171—The parody of Tolkein's *The Lord of the Rings,* originally published in 1969, is The Harvard Lampoon, *Bored of the Rings* (London: Gollancz, 2003).

PAGE 177—*Destination Moon* is available online at www.youtube.com/watch?v=DsisGSBlQqo.

PAGE 182—President Obama's "been there before" remark comes from his remarks on space exploration in the 21st century speech made at the John F. Kennedy Space Center in Florida, transcribed at Barack Obama (2010). *NASA - President Barack Obama on Space Exploration in the 21st Century.* [online] Nasa.gov. Available at: www.nasa.gov/news/media/trans/obama_ksc_trans.html, accessed February 12, 2015.

PAGE 185—Information on the difficulties and potential of the Moon as a colony from Brian Clegg, *Final Frontier* (New York: St. Martin's Press, 2014), pp. 114–22.

14.
IT'S GOOD TO TALK

PAGE 190—The development of the ansible is described in Ursula Le Guin, *The Dispossessed* (London: Gollancz, 1999).

PAGE 190—Details of the Dirac instant transmitter from James Blish, *The Quincunx of Time* (London: The SF Gateway, 2011).

PAGE 198—For background on superluminal experiments using tunneling, see Brian Clegg, *Light Years* (London: Icon Books, 2015), pp. 239–45.

PAGE 200—More details on Chiao's superluminal experiments can be found in A. Kuzmich, A. Dogarlu, L. J. Wang, P. W. Milonni, and R. Y Chiao, "Signal Velocity, Causality, and Quantum Noise in Superluminal Light Pulse Propagation," *Physical Review Letters* 86 (April 2001): 3925–29.

PAGE 202—For a summary of superluminal tunneling from the Nimtz viewpoint, see Horst Aichmann and Günter Nimtz, "The Superluminal Tunneling Story," at arXiv:1304.3155 [physics.gen-ph].

PAGE 204—Information on quantum entanglement from Brian Clegg, *The God Effect* (New York: St. Martin's Griffin, 2009.

PAGE 205—The details of Nick Herbert's instantaneous quantum entanglement communicator are in David Kaiser, *How the Hippies Saved Physics* (New York: W. W. Norton, 2011), pp. 209–14.

PAGE 206—Information on the entangled cat photograph is from Gabriela Barreto Lemos, Victoria Borish, Garrett D. Cole, Sven Ramelow, Radek Lapkiewicz, and Anton Zeilinger, "Quantum imaging with undetected photons," *Nature* (2014), doi: 10.1038/nature13586.

PAGE 206—Interview with Anton Zeilinger undertaken by the author after a lecture at the University of Cambridge Department of Applied Mathematics and Theoretical Physics in October 2004.

15.

CYBORG

PAGE 209—The first appearance of "cyborg" in print was in *The New York Times,* May 22, 1960, 31.

PAGE 211—Information on cochlear implants and other neuroprosthetics from Brian Clegg, *Upgrade Me* (New York: St. Martin's Press, 2008), pp. 211–36.

PAGE 213—The mapping process for phosphenes using Dobelle's technology is described in Stephen Kotler, "Vision Quest," *Wired* 10 (2002): 94–101 .

PAGE 215—The paper describing the experiment giving direct control from the brain of electronic devices is L. R. Hochberg et al, "Neuronal ensemble control of prosthetic devices by a human with tetraplegia," *Nature* 442 (2006): 164–71.

PAGE 217—The experiment using monkeys' premotor cortex to predict intended motion is described in G. Santhanam et al, "A high-performance brain-computer interface" *Nature* 442 (2006): 195–98.

PAGE 218—The one-year experiment with a prosthetic arm using electrodes implanted in the patient's arm is described in Max Ortiz-Catalan, Bo Håkansson, and Rickard Brånemark, "An osseointegrated human-machine gateway for long-term sensory feedback and motor control of artificial limbs," *Science Transitional Medicine* 6 (October 8, 2014): 257.

PAGE 219—Details of the remote-controlled cyborg beetle from Emily Anthes, *Frankenstein's Cat: Cuddling up to Biotech's Brave New Beasts* (Oxford, UK: One World, 2013), pp. 146–66.

PAGE 220—Information on Backyard Brains's cyber cockroach from their Web site www.backyardbrains.com, accessed September 9, 2014.

PAGE 222—More details of Kevin Warwick's "cyborg" experiments are available from his Web site www.kevinwarwick.com, accessed September 9, 2014.

PAGE 224—The 2013 human-rat cyber-telepathy is described in Sara Reardon, "Interspecies telepathy: human thoughts make rat move," *New Scientist,* April 3, 2013, accessed September 8, 2014, www.newscientist.com/article/dn23343-interspecies-telepathy-human-thoughts-make-rat-move.html.

PAGE 224—The 2014 human cyber-telepathy is described in Carles Grau, Romuald Ginhoux et al, "Conscious Brain-to-Brain Communication in Humans Using Non-Invasive Technologies," *PLOS ONE,* accessed August 19, 2014, doi: 10.1371/journal.pone.0105225.

16.

ENGAGE CLOAKING DEVICE

PAGE 226—The quote describing how a human is almost transparent is from H. G. Wells, *The Invisible Man* (London: Penguin Classics, 2012 Kindle edition), chapter 19

PAGE 228—The suggestion that the complexity of magic spells was to make them an entry point into a secret fraternity is from Philip Ball, *Invisible* (London: Bodley Head, 2014), p. 27.

PAGE 229—The way that chameleons use skin coloration for temperature regulation and signaling rather than concealment is described in Philip Ball, *Invisible* (London: Bodley Head, 2014), p. 233.

PAGE 230—The evolution of the peppered moth's concealment and variation when its habitat was darkened by industrial pollution is covered in Mark Fellowes and Nicholas Battey, *30-Second Evolution* (London: Icon Books, 2015), p. 104.

PAGE 231—Gene Roddenberry's indication that cloaking devices were too sneaky comes from "Memory Alpha," The Star Trek Wiki, accessed July 28, 2014, en.memory-alpha.org/wiki/Cloaking_device.

PAGE 237—The BAE Systems tank invisibility shield making it appear to be a family car is described in Peter Forbes and Tom Grimsey, *Nanoscience* (Winterbourne, UK : Papadakis, 2014), p. 94.

17.

OPEN THE POD BAY DOORS

PAGE 241—Marvin Minsky's near miss with a wrench on the set of *2001* and prediction of Hal-like abilities in "four and four hundred years" is in an interview with David Stork in David G. Stork (ed.), *Hal's Legacy* (Cambridge, MA: MIT Press, 2000), p. 24.

PAGE 242—Arthur C. Clarke's explicit denial that Hal was based on IBM by shifting each letter by one position in the alphabet is in Arthur C. Clarke's foreword to David G. Stork (ed.), *Hal's Legacy* (Cambridge, MA: MIT Press, 2000), p. xi.

PAGE 244—Murray Campbell's examination of Hal's chess game is in his section, "'An Enjoyable Game': How HAL Plays Chess," in David G. Stork (ed.), *Hal's Legacy* (Cambridge, MA: MIT Press, 2000), pp. 75–98.

PAGE 244—The history of the chess-playing mechanical Turk is described in Tom Standage, *The Mechanical Turk: The True Story of the Chess Playing Machine That Fooled the World* (London: Penguin Books, 2003).

PAGE 246—Information for the summary of early chess programs from the Murray Campbell section, "'An Enjoyable Game': How HAL Plays Chess," in David G. Stork (ed.), *Hal's Legacy* (Cambridge, MA: MIT Press, 2000), pp. 83–85.

PAGE 248—Information on Roger Bacon from Brian Clegg, *Roger Bacon: The First Scientist* (London: Constable, 2013).

PAGE 249—Quote about Bacon's talking head is from Anon, *History of Friar Bacon: Containing the Wonderful Things That He Did in His Life—Also the Manner of His Death, with the Lives and Deaths of the Two Conjurers, Bungye and Vandermast* (London: Banton, 1991), p. 17.

PAGE 249—Information on the history of speaking machines from the Joseph P. Olive section "'The Talking Computer': Text to Speech Synthesis" in David G. Stork (ed.), *Hal's Legacy* (Cambridge, MA: MIT Press, 2000), pp. 103–4.

PAGE 249—Internet video of dog saying "sausages" on YouTube at www.youtube.com/watch?v=ajsCY8SjJ1Y, accessed September 3, 2014.

PAGE 250—The first computer music, produced at the University of Manchester, is described in B. Jack Copeland, *Turing: Pioneer of the Information Age* (Oxford, UK: Oxford University Press, 2012), pp. 163–64.

PAGE 253—Information in the history of speech recognition from the Raymond Kurzweil section, "When will HAL Understand what we are Saying? Computer Speech Recognition and Understanding," in David G. Stork (ed.), *Hal's Legacy* (Cambridge, MA: MIT Press, 2000), pp. 145–50.

PAGE 254—Apple's Knowledge Navigator appears at a number of locations on YouTube including www.youtube.com/watch?v=QRH8eimU_20, accessed September 3, 2014.

PAGE 256—The claim that the Eugene Goostman chatbot passed the Turing Test is described in BBC News, "Computer AI passes Turing test in 'world first,'" accessed September 2, 2014, at www.bbc.co.uk/news/technology-27762088.

PAGE 261—The arguments that Hal isn't really intelligent are from the Douglas B. Lenat section, "From *2001* to 2001: Common Sense and the Mind of HAL" in David G. Stork (ed.), *Hal's Legacy* (Cambridge, MA: MIT Press, 2000), pp. 193–94.

PAGE 263—The novel featuring a Delphi-based government is John Brunner, *Shockwave Rider* (London: Dent, 1975).

PAGE 263—The work of futurology that inspired *Shockwave Rider* is Alvin Toffler, *Future Shock* (London: Pan Books, 1970).

18.

NEVER-ENDING STORY

PAGE 270—Information on space elevators and the difficulties in constructing them from Brian Clegg, *Final Frontier* (New York: St. Martin's Press, 2014), pp. 61–65.

INDEX